Advances in Time Series Forecasting

(Volume 2)

Edited by

Cagdas Hakan Aladag

Department of Mechanical and Industrial Engineering,
University of Toronto, Canada

Department of Statistics, Hacettepe University, Turkey

General:

1. Any dispute or claim arising out of or in connection with this License Agreement or the Work (including non-contractual disputes or claims) will be governed by and construed in accordance with the laws of the U.A.E. as applied in the Emirate of Dubai. Each party agrees that the courts of the Emirate of Dubai shall have exclusive jurisdiction to settle any dispute or claim arising out of or in connection with this License Agreement or the Work (including non-contractual disputes or claims).
2. Your rights under this License Agreement will automatically terminate without notice and without the need for a court order if at any point you breach any terms of this License Agreement. In no event will any delay or failure by Bentham Science Publishers in enforcing your compliance with this License Agreement constitute a waiver of any of its rights.
3. You acknowledge that you have read this License Agreement, and agree to be bound by its terms and conditions. To the extent that any other terms and conditions presented on any website of Bentham Science Publishers conflict with, or are inconsistent with, the terms and conditions set out in this License Agreement, you acknowledge that the terms and conditions set out in this License Agreement shall prevail.

Bentham Science Publishers Ltd.
Executive Suite Y - 2
PO Box 7917, Saif Zone
Sharjah, U.A.E.
Email: subscriptions@benthamscience.org

**BENTHAM
SCIENCE**

CONTENTS

PREFACE

Human interest in the future can be traced back to prehistoric times. People have always wanted to see what can happen in the future. The future is unknown and mysterious. People have always tried to solve the mystery of the future by using different ways for profit, fame, power or just curiosity sake. Today forecasting is a multibillion dollar industry. All economic publications publish many economic forecasting studies; political writers proclaim on political trends and forthcoming government policies; stockbrokers and financial experts predict stock market trends, when to buy, and what stocks to choose; and many other examples can be given which have application to other fields.

Before making plans or making decisions, an estimate must be made of what conditions will exist over some future period. It is a well-known fact that there is uncertainty about the future. In order to predict to future by dealing with this uncertainty, forecasting is performed. At the present time, forecasting is a challenge which has to be overcome in many fields of application. Forecasting can be considered as a process of using various tools and techniques. Many methods for forecasting the future have been proposed in the literature over the past few decades because of the importance of this popular topic. One way to forecast the future is to use time series analysis. There have been many time series forecasting approaches in the literature. It is possible to divide these approaches into two subclasses which are conventional and advanced forecasting methods. Since conventional approaches such as Box-Jenkins methods has some restrictions such as some assumptions, they cannot always produce reliable forecasts for real world time series. Furthermore, conventional approaches cannot model some real world time series because of the specific characteristic of data. Advanced methods such as neural networks, fuzzy time series, or hybrid approaches have been recently used in many applications in order to deal with these restrictions arising from conventional methods and to get more reliable forecasts. Most of the time, these approaches have been competed to each other. On the other hand, it should be noted that these approaches are complementary rather than competitive. For example, hybrid approaches are very effective forecasting tools. And, these approaches sometimes combine conventional and advanced forecasting methods.

The book intends to be a valuable source of recent knowledge about advanced time series forecasting techniques. New capable advanced forecasting frameworks are discussed and their applicability is shown. The book includes applications of some powerful recent forecasting approaches to real world time series. Besides recent advanced forecasting methods, new efficient forecasting methods are firstly introduced in the book. The readers can find useful information about advanced time series forecasting, as well as its application to real-life problems in various domains. I hope the materials covered in this book, provided by the respectful scholars in the field, motivate and accelerate future progress and introduce new branches off the time series forecasting.

In Chapter 1, Aladag and Turksen have introduced a new performance measure by defining a novel distance measure to evaluate forecasting performance of fuzzy time series. Bas and Egrioglu, in Chapter 2, have suggested a novel fuzzy time series forecasting approach that has a network structure. In Chapter 3, Zarandi et al. has discussed some Type-1 and Type-2 fuzzy time series forecasting models. Chapter 4, by Egrioglu et al., introduce a new neural network model including deterministic trend and seasonality components. In Chapter 5, Yolcu has presented a fuzzy time series method based on genetic algorithms. Aladag and Guney, in Chapter 6, have applied a fuzzy time series forecasting model based on Markov chain transition matrix to stock exchanges. In Cahapter 7, Yolcu has proposed a new high order multivariate fuzzy time series forecasting model. Chapter 8, by Dalar et al., has discussed a framework for using fuzzy functions in fuzzy time series forecasting. Sarica et al., in Chapter 9, have introduced Recurrent ANFIS model for time series forecasting. In Chapter 10, Bas has proposed a hybrid forecasting approach which combines genetic algorithms, differential evolution algorithms, and fuzzy time series.

The editor would also like to express his sincere thanks to all authors for their valuable contributions. The editor would also like to acknowledge valuable assistance from Shehzad Naqvi from Bentham Science Publishers.

Dr. Cagdas Hakan Aladag

Knowledge/Intelligence Systems Laboratory,
Department of Mechanical and Industrial Engineering,
University of Toronto, Toronto,
Canada

Department of Statistics,
Faculty of Science,
Hacettepe University, Ankara,
Turkey

List of Contributors

Ali Z. Dalar Department of Statistics, Faculty of Arts and Sciences, Giresun University, Giresun, Turkey

Barış Asıkgil Department of Statistics, Faculty of Arts and Science, Marmara University, İstanbul, Turkey

Busenur Sarıca Department of Statistics, Faculty of Arts and Science, Marmara University, İstanbul, Turkey

Cagdas Hakan Aladag Knowledge/Intelligence Systems Laboratory, Department of Mechanical and Industrial Engineering, University of Toronto, Toronto, Canada
Department of Statistics, Faculty of Science, Hacettepe University, Ankara, Turkey

Eren Bas Department of Statistics, Faculty of Arts and Sciences, Giresun University, Giresun, Turkey

Erol Egrioglu Department of Statistics, Faculty of Arts and Sciences, Giresun University, Giresun, Turkey

Hilal Guney Department of Statistics, Gazi University, Ankara, Turkey

I. Burhan Turksen Department of Industrial Engineering, TOBB University of Economics and Technology, Ankara, Turkey
Department of Mechanical and Industrial Engineering, University of Toronto, Toronto, Canada

M. Yalinezhaad Department of Industrial Engineering and Management Systems, Amirkabir University of Technology (Polytechnic of Tehran), Tehran, Iran

M.H. Fazel Zarandi Department of Industrial Engineering and Management Systems, Amirkabir University of Technology (Polytechnic of Tehran), Tehran, Iran

Ozge Cagcag Yolcu Department of Industrial Engineering, Faculty of Engineering, Giresun University, Giresun, Turkey

Ufuk Yolcu Department of Statistics, Faculty of Science, Ankara University, Ankara, Turkey
Department of Econometrics, Faculty of Economics and Administrative Sciences, Giresun University, Giresun, Turkey

iv

Advances in Time Series Forecasting

Volume # 2

Editor: Cagdas Hakan Aladag

ISSN (Online): 2543-280X

ISSN (Print): 2543-2796

ISBN (Online): 978-1-68108-528-9

ISBN (Print): 978-1-68108-529-6

©2017, Bentham eBooks Imprint.

Published By Bentham Science Publishers – Sharjah, UAE. All Rights Reserved.

CHAPTER 1

Fuzzy Time Series Forecasting Models Evaluation Based on A Novel Distance Measure

Cagdas Hakan Aladag[1,*] and **I. Burhan Turksen[2]**

[1] *Department of Mechanical and Industrial Engineering, University of Toronto, Toronto, Canada*

[2] *Department of Industrial Engineering, TOBB University of Economics and Technology, Ankara, Turkey*

Abstract: In the literature, many models based on fuzzy systems have been utilized to solve various real world problems from different application areas. One of this areas is time series forecasting. Successful forecasting results have been obtained from fuzzy time series forecasting models in many studies. To determine the best fuzzy time series model among possible forecasting models is a vital decision. In order to evaluate fuzzy time series forecasting models, conventional performance measures such as root mean square error or mean absolute percentage error have been widely utilized in the literature. However, the nature of fuzzy logic is not taking into consideration when such conventional criteria are employed since these criteria are computed over crisp values. When fuzzy time series forecasting models are evaluated, using criteria which work based on fuzzy logic characteristics is wiser. Therefore, Aladag and Turksen [2] suggested a new performance measure which is calculated based on membership values to evaluate fuzzy systems. It is called as membership value based performance measure. In this study, a novel distance measure is firstly defined and a new membership value based performance measure based on this new distance measure is proposed. The proposed criterion is also applied to real world time series in order to show the applicability of the suggested measure.

Keywords: Forecasting, Fuzzy time series, Membership value based performance measure, Membership values, Model evaluation, Performance criterion, Real world time serie.

INTRODUCTION

There have been many different fuzzy logic based models to solve real world problems in the literature. Sometimes, it is possible to use more than one model for a given problem. In this case, determining the best model is a vital decision. To compare different models, performances of these models ought to be

---

[*] **Corresponding author Cagdas Hakan Aladag**: Department of Mechanical and Industrial Engineering, University of Toronto, Toronto, Canada; E-mail: chaladag@gmail.com

measured. Therefore, to evaluate such models is an important issue. In the literature, there are many studies in which a performance measure is used to evaluate different forecasting models based on fuzzy logic. Some of these studies are Aladag [1]; Aladag and Turksen [2]; Avazbeigi *et al.* [7]; Cai *et al.* [10]; Chen *et al.* [11]; Chen and Chen [12]; Chen and Chen [13]; Cheng and Li [15]; Chen and Kao [14]; Egrioglu *et al.* [16]; Lee and Hong [21]; Li *et al.* [22]; Lu *et al.* [23]; Singh and Borah [29]; Wang [33]; Yolcu *et al.* [34].

In general, fuzzy systems are composed of three fundamental stages such as fuzzification, fuzzy inference, and defuzzification [8]. In other words, computing the output of fuzzy systems usually passes through three stages which are fuzzification, fuzzy inference, and defuzzification [18]. In most of the fuzzy systems, fuzzy outputs of the system are obtained after fuzzy inference is performed by using membership values. Then, fuzzy outputs are defuzzified using a proper method. Finally, a performance measure is calculated based on the difference between the defuzzified outputs and the desired values. That is, the performance measure is computed using crisp values. Thus, the membership values are not taken into consideration. It is because membership values carry important information that evaluating the performance of a fuzzy logic based model with ignoring the membership values will lead to misleading results. Since fuzzy inference is performed using membership values, these membership values should also be employed to compute a performance criterion. In such a case, it is unnecessary to defuzzify fuzzy outputs. Thus, using a performance measure which utilizes the membership values would be wiser.

In this study, a new performance measures based on the membership values is improved to evaluate fuzzy logic based systems. A new distance between outputs and targets is defined. By using this new distance, a new performance measures is proposed. There are many fuzzy systems in the literature. In this study, we focus on fuzzy time series. However, the proposed measure can be easily used for other fuzzy systems since it is calculated based on the membership values. In order to explain the suggested measure better and to show the applicability of it, we utilize fuzzy time series forecasting models.

In the fuzzy time series literature, the outputs and the desired values are the predicted and corresponding observations, and a performance measure is generally computed without taking into consideration the membership values of both predictions and observations. In general, a performance criterion such as mean square error (MSE), root mean square error (RMSE) or mean absolute percentage error (MAPE) which is calculated using crisp values with ignoring the membership values has been used in the literature. However, the membership values should be employed since these values carry important information.

Ignoring the membership values can lead to misleading evaluation results. There have been various performance criteria used in fuzzy time series to measure the performance of models. On the other hand, a performance criterion which takes into consideration the membership values is just one proposed by Aladag and Turksen [2]. This proposed performance measure was called membership value based performance measure (MPM). Lack of performance measures based on the membership values is a big gap in the fuzzy time series literature. Also, evaluation of models based on fuzzy systems is a vital subject in this area. It is obvious that this gap should be scientifically filled.

Forecasting is a popular research topic that is attracting more and more attention from researchers and practitioners in various fields [20]. An important issue is to determine the best forecasting model which gives the most accurate results. In order to determine the best model, one needs to evaluate the forecasting performance of various models. In the literature, various performance measures have been utilized to determine the best forecasting model [19, 24, 26]. Armstrong and Collopy [6] performed a comparison study to evaluate some forecasting performance measures. Shcherbakov *et al.* [28] also reviewed different forecasting performance measures in their survey study. In recent years, forecasting models based on fuzzy logic and artificial intelligent methods have been employed in order to get more accurate forecasts [25, 35, 36]. Among these, fuzzy time series forecasting models are the most widely used ones for time series that contains uncertainty [4, 5]. And, various fuzzy time series models have been proposed in the literature. In many studies available in the fuzzy time series literature, a performance criterion such as MSE, RMSE, or MAPE and so forth has been utilized to evaluate the performance of fuzzy time series forecasting models.

Using a performance criterion such as MSE, RMSE, or MAPE which are calculated by using crisp values can also bring some disadvantages when such measures are tried to be used for fuzzy logic based models. These disadvantages can be given as follows [2]:

"When a performance measure such as RMSE, MSE or MAPE is applied, it will be necessary to perform a defuzzification process. If defuzzification phase is performed, an error arises since a fuzzy prediction is tried to be mapped into a crisp value. The total prediction error of a model containing defuzzification phase is composed of (i) the forecasting method and (ii) the method for defuzzification. Hence, the total error of a model can be decreased if defuzzification is not performed. In addition, in the literature, there are different methods for defuzzification. Even for same fuzzy prediction process, different crisp values can be obtained from different defuzzification methods. This means that the value of a conventional performance criterion such as MSE, RMSE or MAPE will change depending on the method used for a particular defuzzification method. This will lead to both inconsistent forecasting results and inconsistent evaluation results.

In addition to disadvantages addressed above, the performance criteria such as RMSE or MAPE which are calculated over numerical values cannot be used when decision makers have to use linguistic variables. The concept of linguistic variable was firstly used by Zadeh [38] to handle the approximate reasoning. Sometimes, both the inputs and the outputs of a fuzzy system are linguistic terms [9]. In such a case, to provide support to the decision makers in the process of making a choice among different options, we suggest an alternative performance measure which utilizes the membership values. Already, in fuzzy time series, if the researcher does not look for crisp forecast values, using another method for defuzzification would not be necessary after the fuzzy predictions are obtained."

Fuzzy set theory introduced by Zadeh [37] was firstly adopted in time series by Song and Chissom [30 - 32] to deal with uncertainty. And, the approach was called as fuzzy time series. Following Song and Chissom's fuzzy time series model, many fuzzy time series models have been proposed for forecasting [27]. As mentioned above, to evaluate a fuzzy logic based model, a performance measure such as RMSE, MSE or MAPE is calculated based on the difference between the outputs of the model and the corresponding desired values. This calculation is performed over crisp values even though fuzzy inference is performed by using membership values. Using such a performance measures can also bring some disadvantages. Therefore, Aladag and Turksen [2] proposed a new performance measure in which membership values are used to calculate a performance measure. In this study, the process of evaluation of fuzzy time series models was examined, the approach proposed by Aladag and Turksen [2] was extended and a new kind of performance measure that utilize membership values was also be proposed in order to determine more accurate fuzzy time series forecasting models. Also, the new performance measure proposed in this research is applied to three real world time series which are index 100 in stocks and bonds exchange market of İstanbul, the number of people who die in traffic accidents in Turkey, and the enrolment data of Alabama University which is a well-known data in fuzzy time series literature.

In this chapter, the proposed distance measure and the suggested performance criterion based on this distance measure are introduced in the next section. The implementation and the obtained results are presented in section whose title is the application. Finally, the last section concludes the chapter.

THE PROPOSED DISTANCE MEASURE AND THE SUGGESTED PERFORMANCE CRITERION

Let a be the obtained prediction for b which represents an observation. In any other fuzzy system study, a would be the output and b would be the corresponding desired or target value. Let A and B be the vectors whose elements are the membership values of prediction a and corresponding observation b, respectively. A and B vectors are as follows:

A = [0.003 0.010 0.044 0.120 0.175 0.648] (for prediction a)

B = [0.003 0.122 0.021 0.122 0.575 0.157] (for observation b)

It is clearly seen that the number of clusters is 6 for this example since vectors have six elements. In this representation, each element represents a membership value for a corresponding cluster. For example, 0.010 and 0.648 are the degrees of belongingness of prediction a to second cluster and to sixth cluster, respectively. Definitions of these clusters depend on the nature of data. For instance, if observations are temperatures, these cluster 1, 2, 3, 4, 5 and 6 can represent "very low", "low", "moderate", "high", "very high", and "extremely high", respectively.

We would like to note that number of clusters is not determined in the process of the proposed performance measure like in all criteria available in the literature. The number of clusters is an input for the proposed MPM. The membership value of a for cluster 1 is 0.003 and this degree for b is also 0.003 so the membership values of a and b are equal. This is a good sign which shows that there is no difference between the prediction and the observation in terms of membership values for cluster 1. On the other hand, the membership value of a for cluster 5 is 0.175 while this degree for b is 0.575. Thus, there is a difference between these membership values. This indicates that there is a difference between the prediction and the observation. It is desired that there is no differences between all mutually corresponding membership values. If all mutually corresponding membership values are very close to each other, it can be said that a is an accurate forecast for the observation b. The less the difference between memberships is, the better the accuracy is. Thus, the proposed distance measures the difference between the membership values.

An anomaly that would result from 0-1 membership values could arise when the distance between these values are calculated using conventional measures. It would be shown that an anomaly would arise if a conventional distance such as Euclidean distance is considered. The following example is given in order to demonstrate our concern. Let o_1 and o_2 be two different outputs for the same corresponding desired value d. Let Ao_1, Ao_2, and Ad be the vectors whose elements are the membership values of predictions o_1 and o_2 and corresponding observation d, respectively. These vectors are given below.

$Ao_1 = [1\ 0\ 0\ 0\ 0\ 0]$, $Ao_2 = [0\ 0\ 0\ 0\ 1\ 0]$, $Ad = [0\ 0\ 0\ 0\ 0\ 1]$

Euclidean distance between Ad and Ao_1 is equal to the distance obtained from Ad and Ao_2. According to this result, two predictions o_1 and o_2 are same for observation d. However, it is obvious that the prediction o_2 is better than o_1 especially when these 6 clusters represent ordinal linguistic variables. The observation d belongs to the last cluster with the maximum membership value. While output o_1 belongs to the first cluster, output o_2 belongs to cluster 5. In this case, it is clear that d is closer to prediction o_2 than is o_1. Therefore, the suggested distance does not only measure the differences between mutually corresponding membership values but also take into consideration cluster orders. The reason is that nature of fuzzy sets should be taken into consideration when a difference between membership values is computed.

For vectors A and B given above, the proposed distance for the suggested criterion is calculated as follows. Two new vectors whose elements are indices are generated. These indices are determined by ordering the membership values. While the minimum index corresponds to maximum membership value, the maximum index corresponds to minimum membership value. For A and B, these new vectors As and Bs are generated as follows:

$A = [0.003\ 0.010\ 0.044\ 0.120\ 0.175\ 0.648]$ ➜ $As = [6, 5, 4, 3, 2, 1]$

$B = [0.003\ 0.122\ 0.021\ 0.122\ 0.575\ 0.157]$ ➜ $Bs = [6, 3 - 5, 3, 4, 1, 2]$

Since B includes a repeated value (0.122), corresponding elements in Bs is adjusted by using mean for this value. Thus, As and Bs can be written as follows:

$As = [6, 5, 4, 3, 2, 1]$ ➜ $As = [6, 5, 4, 3, 2, 1]$

$Bs = [6, 3 - 5, 3, 4, 1, 2]$ ➜ $Bs = [6\ 3.5\ 5\ 3.5\ 1\ 2]$

Then, the distances for each membership are calculated by taking into account both the difference between mutually corresponding membership values and the cluster orders. Let A and B keep membership values for an observation and a prediction, respectively. The formula for calculation i^{th} distance can be given as follows:

$$d_i = |A_i - B_i| / A_i * |As_i - Bs_i| / (cn - 1), i=1,2, ..., cn \qquad (1)$$

where A_i, B_i, As_i, and Bs_i are i^{th} elements of vectors A, B, As, and Bs, respectively, and cn is the number of clusters. $1/(cn - 1)$ term is used to rescale the obtained value in accordance with the structure of the proposed measure. For A and B, all computations are presented in Table **1**. For the given example, sum of membership values A_i and B_i equals to 1 but it is not supposed that the sum of membership values equals to 1. The sum of membership values can equal to 1 or not. In both cases, the proposed criterion can be easily calculated. If it is necessary that the sum of membership values equals to 1 because of the nature of used fuzzy logic based system then membership values can be easily adjusted so that sum of the membership values equals to 1. This case will also be explained further on in this section.

Table 1. Calculations for A and B vectors.

A	As	B	Bs	Formula	d_i
0.003	6	0.003	6	|0.003-0.003|/0.003 * |6-6|/(6-1)	0.0000
0.010	5	0.122	3.5	|0.010-0.122|/0.010 * |5-3.5|/(6-1)	3.3600
0.044	4	0.021	5	|0.044-0.021|/0.044 * |4-5|/(6-1)	0.1045
0.120	3	0.122	3.5	|0.120-0.122|/0.120 * |3-3.5|/(6-1)	0.0017
0.175	2	0.575	1	|0.175-0.575|/0.175 * |2-1|/(6-1)	0.4571
0.648	1	0.157	2	|0.648-0.157|/0.648 * |1-2|/(6-1)	0.1515

The proposed distance for A and B vectors is the mean value of all computed distances. Thus, the proposed distance can be calculated as follows:

$$\frac{1}{cn}\sum_{i=1}^{cn} d_i = \frac{(0 + 3.36 + 0.1045 + 0.0017 + 0.4571 + 0.1515)}{6} = 0.6791$$

The proposed distance between A and B is equal to 0.6791. The lesser this mean value is, the closer the output and the desired value are. The proposed distance value above is calculated based on the differences between not only mutually corresponding membership values but also indices represent cluster orders.

Why is such a distance proposed? This is a key point. Let's examine the formula of the suggested distance in (1). The formula is composed of two main parts. The first part is as follows:

$$|A_i - B_i| \,/\, A_i$$

This part calculates the distance between membership values as a percentage.

Therefore, the error deviation can be calculated as a percentage. For instance, 0.00001 can be used for any A_i which equals to 0. And the second part of the formula is given below.

$$|As_i - Bs_i| / (cn - 1)$$

The second part calculates distance between indices. As mentioned before, cluster orders carry important information because of the characteristic of fuzzy logic. Therefore, it is important that how close the indices in vectors of observations and corresponding predictions. If mutually corresponding indices are same, the obtained prediction is accurate. Finally, by multiplying these two parts, the suggested distance is obtained.

As seen from the first part of the formula, if A_i is 0, then the proposed distance will be undefined. In order to deal with this problem, if A_i is 0, a very small value which is close to 0 but not equals to 0 can be used for this A_i. Another anomaly could arise when the difference between As_i and Bs_i equals to 0. In this case, the distance will be computed as 0. On the other hand, there can be a distance because of the first part of the formula. If the distance is computed as 0 in such a case, then the first part of the formula cannot be taken into account. Not to diminish the effect of the first part of the formula, the difference between As_i and Bs_i can be taken as 0.01 when this difference equals to 0. Hence, the effect of the first part of the formula is taken into consideration even if the difference between As_i and Bs_i is 0.

Sometimes, observations can be fuzzified by using fuzzy c-means method so the sum of the membership values of any observation is 1. On the other hand, the fuzzy time series approach which will be evaluated can produce membership values whose sum does not equal to 1 for any observation even though fuzzy c-means method was employed for fuzzification. In such a case, the membership values obtained from the fuzzy time series approach for each observation are adjusted so that sum of the membership values equals to 1. For example, let Ao and Ad be vectors whose elements are the membership values of a prediction and a corresponding observation, respectively. Ao and Ad are as follows:

Ad = [0.12 0.01 0.24 0.10 0.04 0.49], Ao = [0.34 0.14 0.09 0.10 0.16 0.88]

Although sum of the membership values of Ad is 1, sum of the membership values of Ao is 1.71. Thus, in order to calculate the difference between Ao and Ad according to MPM, the membership values included in Ao are adjusted such that sum of the membership values equals to 1. This can be done by using the simple formula given below.

$$Ao'_i = \frac{1}{S}Ao_i, i = 1,2, \dots, cn \qquad (2)$$

where Ao' represents the adjusted membership values of the prediction and S and cn are sum of the membership values and the number of clusters, respectively. For given example, S and cn are 1.71 and 6, respectively. Thus, by using the formula (2), vector Ao' is obtained as follows:

Ao' = [0.19883 0.08187 0.05263 0.05848 0.09357 0.51462]

Like in Ad, sum of the membership values of Ao' is 1 now. Then, the distance between Ad and Ao' can be calculated by using the formula (1) and the distance value is obtained as 0.0268.

After the distance measure proposed in this study is defined, the suggested performance criterion based on this new distance measure can be explained. The MPM proposed by Aladag and Turksen [2] is extended by using this new distance. Then, a new measure is proposed to evaluate fuzzy logic based systems. The proposed measure can be computed as follows:

The distance value calculated by using the formula (1) can be used to measure the only difference between one prediction and a corresponding observation. On the other hand, in fuzzy time series, more than one prediction and corresponding observation are generally employed to evaluate a fuzzy time series approach. For instance, a time series can be split into training and test sets and the test set can be used for evaluation. A fuzzy time series approach is run over the training set and inference is performed. After this process, fuzzy forecasts for test set are obtained. Then, a performance criterion is calculated based on the difference the obtained predictions and corresponding observations in the test set. If the test set has T observations, the suggested measure will be calculated over the test set by using the formula given in (3).

$$\frac{1}{T}\sum_{j=1}^{T} distance_j \qquad (3)$$

As seen from the formula above, the proposed measure is the mean of distances which is obtained by using the formula (1). For T observations included in test set, the proposed criterion is obtained by calculating T distances for T pairs of predictions and corresponding observations. When all predictions and corresponding observations are exactly the same, the proposed measure is 0. Therefore, the closer to 0 the proposed measure value is, the more accurate the predictions are.

THE APPLICATION

To show the applicability of the proposed performance measure, three different real world time series are used in the implementation. These series are index 100 in stocks and bonds exchange market of İstanbul (IMKB100) and the enrolment data of Alabama University which is a well-known data in fuzzy time series literature. Two different fuzzy time series forecasting approaches which are proposed by Aladag *et al.* [3] and Egrioglu [17] are applied to these time series. And, the performances of these fuzzy logic based approaches are evaluated by using the proposed performance measure.

In order to clearly explain how the proposed performance measure works, all computations for the enrolment data will be given in detail. First of all, to apply these methods to the enrollment data, the membership values of observations were obtained by using fuzzy c-means clustering method. Because of the nature of the data, the number of clusters was taken as 7. The enrollment data and the membership values of observations are presented in Table **2**.

Table 2. The enrollment data and the membership values of observations.

Years	Observ.	Cluster 1	Cluster 2	Cluster 3	Cluster 4	Cluster 5	Cluster 6	Cluster 7
1971	13055	0.8888	0.0426	0.0279	0.0190	0.0112	0.0062	0.0043
1972	13563	0.9925	0.0033	0.0019	0.0012	0.0006	0.0003	0.0002
1973	13867	0.8062	0.0941	0.0478	0.0274	0.0137	0.0065	0.0043
1974	14696	0.0474	0.7630	0.1216	0.0435	0.0153	0.0058	0.0035
1975	15460	0.0000	0.0001	0.9997	0.0001	0.0000	0.0000	0.0000
1976	15311	0.0048	0.1654	0.7821	0.0377	0.0069	0.0020	0.0011
1977	15603	0.0039	0.0482	0.7981	0.1342	0.0115	0.0027	0.0014
1978	15861	0.0017	0.0133	0.0598	0.9121	0.0104	0.0019	0.0009
1979	16807	0.0001	0.0003	0.0005	0.0013	0.9971	0.0005	0.0002
1980	16919	0.0006	0.0018	0.0030	0.0071	0.9821	0.0042	0.0013
1981	16388	0.0093	0.0406	0.0899	0.4379	0.3871	0.0249	0.0103
1982	15433	0.0001	0.0025	0.9955	0.0016	0.0002	0.0001	0.0000
1983	15497	0.0004	0.0071	0.9830	0.0081	0.0010	0.0003	0.0001
1984	15145	0.0061	0.7852	0.1751	0.0250	0.0059	0.0018	0.0011
1985	15163	0.0069	0.7232	0.2293	0.0303	0.0070	0.0022	0.0012
1986	15984	0.0001	0.0004	0.0013	0.9977	0.0005	0.0001	0.0000
1987	16859	0.0000	0.0001	0.0002	0.0006	0.9987	0.0003	0.0001
1988	18150	0.0000	0.0000	0.0000	0.0000	0.0001	0.9997	0.0001

(Table 2) contd.....

Years	Observ.	Cluster 1	Cluster 2	Cluster 3	Cluster 4	Cluster 5	Cluster 6	Cluster 7
1989	18970	0.0010	0.0018	0.0023	0.0032	0.0063	0.0438	0.9416
1990	19328	0.0010	0.0017	0.0022	0.0029	0.0052	0.0239	0.9632
1991	19337	0.0010	0.0019	0.0024	0.0031	0.0057	0.0258	0.9601
1992	18876	0.0021	0.0041	0.0052	0.0072	0.0147	0.1196	0.8472

In this application, a test set was not used so all observations except from the first one were employed for evaluation. The first observation for year 1971 was not used since two approaches applied to the data are first order approaches. When the fuzzy time series forecasting approach introduced by Aladag *et al.* [3] is applied to the enrollment data, the obtained membership values for forecasts are shown in Table **3**. As seen from Table **3**, the membership values of the prediction for year 1971 is absent in this table since the method proposed by Aladag *et al.* [3] is a first order fuzzy time series approach.

Table 3. The membership values of predictions obtained from the method proposed by Aladag *et al.* [3].

Years	Cluster 1	Cluster 2	Cluster 3	Cluster 4	Cluster 5	Cluster 6	Cluster 7
1972	0.8888	0.0328	0.8126	0.0426	0.0426	0.0279	0.0062
1973	0.9925	0.0033	0.8126	0.0033	0.0033	0.0145	0.0003
1974	0.8062	0.0478	0.8062	0.0941	0.0941	0.0478	0.0065
1975	0.7630	0.1216	0.7630	0.5310	0.7225	0.1216	0.0058
1976	0.9997	0.9997	0.9997	0.0001	0.0001	0.9997	0.0000
1977	0.7821	0.7821	0.7821	0.1654	0.1654	0.7821	0.0020
1978	0.7981	0.7981	0.7981	0.1342	0.0482	0.7981	0.0027
1979	0.0598	0.0598	0.2108	0.9121	0.0133	0.9121	0.0019
1980	0.7788	0.9426	0.9420	0.0013	0.9971	0.0013	0.0005
1981	0.7788	0.9426	0.9420	0.0071	0.9821	0.0071	0.0042
1982	0.3871	0.3871	0.3871	0.4379	0.3871	0.4379	0.0249
1983	0.9955	0.9955	0.9955	0.0025	0.0025	0.9955	0.0001
1984	0.9830	0.9830	0.9830	0.0081	0.0071	0.9830	0.0003
1985	0.7852	0.1751	0.7852	0.5310	0.7225	0.1751	0.0018
1986	0.7232	0.2293	0.7232	0.5310	0.7225	0.2293	0.0022
1987	0.0013	0.0013	0.2108	0.9977	0.0005	0.9977	0.0001
1988	0.7788	0.9426	0.9420	0.0013	0.9987	0.0006	0.0003
1989	0.0001	0.0001	0.0001	0.0001	0.0001	0.0001	0.9997
1990	0.0063	0.7366	0.0063	0.0032	0.0063	0.4649	0.8187

(Table 3) contd.....

Years	Cluster 1	Cluster 2	Cluster 3	Cluster 4	Cluster 5	Cluster 6	Cluster 7
1991	0.0052	0.7366	0.0052	0.0029	0.0052	0.4649	0.8187
1992	0.0057	0.7366	0.0057	0.0031	0.0057	0.4649	0.8187

As seen from Table **3**, for each observation, sums of the obtained membership values are greater than 1. However, for each observation, sums of the membership values in Table **2** are 1 since fuzzy c-means method was used in the fuzzification process. Thus, first of all, the membership values produced by Aladag *et al.* [3] are adjusted so that sum of the membership values of each prediction is equal to 1. This operation can be done by using the formula (2). The adjusted membership values are presented in Table **4**.

Table 4. The adjusted membership values of predictions for the method proposed by Aladag *et al.* [3].

Years	Cluster 1	Cluster 2	Cluster 3	Cluster 4	Cluster 5	Cluster 6	Cluster 7
1972	0.4795	0.0177	0.4384	0.0230	0.0230	0.0151	0.0033
1973	0.5424	0.0018	0.4441	0.0018	0.0018	0.0079	0.0002
1974	0.4237	0.0251	0.4237	0.0494	0.0494	0.0251	0.0034
1975	0.2519	0.0402	0.2519	0.1753	0.2386	0.0402	0.0019
1976	0.2500	0.2500	0.2500	0.0000	0.0000	0.2500	0.0000
1977	0.2260	0.2260	0.2260	0.0478	0.0478	0.2260	0.0006
1978	0.2363	0.2363	0.2363	0.0397	0.0143	0.2363	0.0008
1979	0.0275	0.0275	0.0971	0.4204	0.0061	0.4204	0.0009
1980	0.2126	0.2573	0.2571	0.0004	0.2722	0.0004	0.0001
1981	0.2126	0.2573	0.2571	0.0019	0.2680	0.0019	0.0011
1982	0.1581	0.1581	0.1581	0.1788	0.1581	0.1788	0.0102
1983	0.2497	0.2497	0.2497	0.0006	0.0006	0.2497	0.0000
1984	0.2490	0.2490	0.2490	0.0020	0.0018	0.2490	0.0001
1985	0.2472	0.0551	0.2472	0.1672	0.2275	0.0551	0.0006
1986	0.2288	0.0726	0.2288	0.1680	0.2286	0.0726	0.0007
1987	0.0006	0.0006	0.0954	0.4516	0.0002	0.4516	0.0000
1988	0.2125	0.2572	0.2571	0.0003	0.2726	0.0002	0.0001
1989	0.0001	0.0001	0.0001	0.0001	0.0001	0.0001	0.9994
1990	0.0031	0.3607	0.0031	0.0016	0.0031	0.2277	0.4009
1991	0.0026	0.3613	0.0026	0.0014	0.0026	0.2280	0.4015
1992	0.0028	0.3610	0.0028	0.0015	0.0028	0.2279	0.4012

According to Table **2**, the membership values of observation for year 1972 are as follows:

[0.9925 0.0033 0.0019 0.0012 0.0006 0.0003 0.0002]

When Table **4** is examined, it is seen that the membership values of prediction for year 1972 are as follows:

[0.4795 0.0177 0.4384 0.0230 0.0230 0.0151 0.0033]

Thus, for year 1972, the membership values of both the observation and the prediction are given below.

	Year	Cluster 1	Cluster 2	Cluster 3	Cluster 4	Cluster 5	Cluster 6	Cluster 7
Observation	1972	0.9925	0.0033	0.0019	0.0012	0.0006	0.0003	0.0002
Prediction	1972	0.4795	0.0177	0.4384	0.0230	0.0230	0.0151	0.0033

When the formula (1) is utilized, the proposed distance between the observation and the prediction for 1972 is calculated as 7.4246. For year 1975, the membership values of both the observation and the prediction are given below.

	Year	Cluster 1	Cluster 2	Cluster 3	Cluster 4	Cluster 5	Cluster 6	Cluster 7
Observation	1975	0.0000	0.0001	0.9997	0.0001	0.0000	0.0000	0.0000
Prediction	1975	0.2519	0.0402	0.2519	0.1753	0.2386	0.0402	0.0019

In a similar way, the distance between the observation and the prediction for 1975 is 2539.8813 when the formula (1) is applied. Thus, it can be said that the method proposed by Aladag *et al.* [3] produced a better prediction for year 1972 than this obtained for year 1975 in terms of the proposed distance measure.

For all 21 years, the obtained distances for 21 pairs of predictions and observations are presented in Table **5**. After all distances given in Table **5** are calculated, the proposed criterion can be calculated by using the formula (3). In this case, T is 21 since there are 21 pairs of predictions and observations. Thus, the proposed performance measure was found as 406.5408 by using the formula (3).

Table 5. The obtained distances for all years when the method proposed by Aladag *et al.* [3] is utilized.

Years	distances	Years	distances	Years	distances
1972	7.425	1979	63.876	1986	429.087
1973	0.308	1980	40.124	1987	82.413
1974	0.538	1981	1.969	1988	4795.859
1975	2539.881	1982	351.203	1989	0.227
1976	13.599	1983	117.388	1990	20.124
1977	10.543	1984	12.720	1991	18.434
1978	19.040	1985	4.166	1992	8.506

The fuzzy time series forecasting method proposed by Egrioglu [17] was applied to the enrollment data, and the membership values for predictions were obtained. Then, these membership values were adjusted by using the formula (2) so that sum of the membership values of each prediction is equal to 1. The adjusted membership values are shown in Table **6**. As seen from Table **6**, the membership values of the prediction for year 1971 is absent in this table since the method proposed by Egrioglu [17] is also a first order fuzzy time series approach.

Table 6. The membership values of predictions obtained from the method proposed by Egrioglu [17].

Years	Cluster 1	Cluster 2	Cluster 3	Cluster 4	Cluster 5	Cluster 6	Cluster 7
1972	0.2625	0.2499	0.2375	0.1153	0.0856	0.0362	0.0130
1973	0.2625	0.2499	0.2375	0.1153	0.0856	0.0362	0.0130
1974	0.2496	0.2496	0.2432	0.1181	0.0877	0.0371	0.0148
1975	0.0973	0.1582	0.2022	0.2022	0.1762	0.1306	0.0333
1976	0.1160	0.2319	0.0831	0.2319	0.1980	0.0739	0.0651
1977	0.1321	0.2050	0.0947	0.2050	0.2050	0.0841	0.0742
1978	0.1297	0.2073	0.0930	0.2073	0.2073	0.0826	0.0728
1979	0.1034	0.1785	0.0690	0.1895	0.2234	0.0147	0.2216
1980	0.0006	0.2385	0.0148	0.0859	0.2639	0.2639	0.1324
1981	0.0019	0.2401	0.0149	0.0865	0.2617	0.2617	0.1332
1982	0.1485	0.1541	0.0991	0.1541	0.1541	0.1362	0.1541
1983	0.1163	0.2313	0.0833	0.2313	0.1984	0.0740	0.0653
1984	0.1169	0.2299	0.0838	0.2299	0.1995	0.0744	0.0656
1985	0.0943	0.1532	0.2064	0.2064	0.1707	0.1265	0.0426
1986	0.0960	0.1560	0.1941	0.1941	0.1738	0.1288	0.0574
1987	0.1015	0.1752	0.0677	0.1861	0.2400	0.0119	0.2175

(Table 6) contd.....

Years	Cluster 1	Cluster 2	Cluster 3	Cluster 4	Cluster 5	Cluster 6	Cluster 7
1988	0.0006	0.2383	0.0148	0.0859	0.2641	0.2641	0.1322
1989	0.0000	0.0000	0.1667	0.1667	0.1667	0.1667	0.3332
1990	0.0027	0.0052	0.0364	0.0364	0.0364	0.4156	0.4672
1991	0.0025	0.0046	0.0209	0.0209	0.0209	0.4379	0.4922
1992	0.0027	0.0050	0.0225	0.0225	0.0225	0.4354	0.4895

For all 21 years, the obtained distances for 21 pairs of predictions and observations are presented in Table **7**. After all distances given in Table **7** are calculated, the proposed criterion can be calculated by using the formula (3). In this case, T is 21 since there are 21 pairs of predictions and observations. According to the membership values of observations given in Table **2** and of predictions presented in Table **6**, the suggested performance measure for the method proposed by Egrioglu [17] was computed as 216.9368 by using the formula (3).

Table 7. The obtained distances for all years when the method proposed by Egrioglu [17] is utilized.

Years	distances	Years	distances	Years	distances
1972	0.861	1979	206.799	1986	9.541
1973	0.030	1980	14.407	1987	409.036
1974	0.206	1981	1.618	1988	2238.451
1975	1061.319	1982	581.130	1989	3.573
1976	1.618	1983	13.692	1990	0.565
1977	1.861	1984	2.086	1991	0.286
1978	5.954	1985	2.541	1992	0.100

The obtained performance measure values for both methods are given in Table **8**. According to Table **8**, the method proposed by Egrioglu [17] gives more accurate forecasts than those obtained from Aladag *et al.* [3] in terms of the proposed performance measure.

Table 8. The obtained proposed performance measure values for both methods.

Method	The proposed criterion
Aladag *et al.* [3]	406.5408
Egrioglu [17]	216.9368

It should be noted that for both methods, after the membership values of predictions are obtained, these predictions are not defuzzified to evaluate the performances of the methods. However, if a conventional performance measure such as MSE, RMSE or MAPE were used, it would be necessary to use a method for defuzzification to obtain crisp values. In the studies Aladag *et al.* [3] and Egrioglu [17], MSE values were used to assess the performance of the suggested models. To use this criterion, the obtained fuzzy forecasts were defuzzified, and then MSE values were calculated over numerical values. MSE values obtained from these methods are presented in Table **9**.

Table 9. MSE values given in Aladag *et al.* [3] and Egrioglu [17].

Method	MSE
Aladag *et al.* [3]	46423.01
Egrioglu [17]	234846.85

According to Table **9**, the method proposed by Aladag *et al.* [3] produces better forecasts than those obtained from Egrioglu [17] in terms of MSE criterion. On the other hand, for this application, Egrioglu's [17] method is better than the method proposed by Aladag *et al.* [3] in terms of the proposed performance measure. Using the proposed criterion computed over membership values or using another conventional measure such as MSE calculated over crisp values is a decision that belongs to decision maker. If a decision maker would like to exploit the advantages of the proposed criterion, he/she can prefer to use the proposed criterion.

In this study, a new performance measure based on membership values is proposed. In the literature, the other performance measure computed over membership values was proposed by Aladag and Turksen [2]. As mentioned before, this criterion was called MPM. For the aim of comparison, the criterion suggested by Aladag and Turksen [2] is also utilized to evaluate these two fuzzy time series approach. All obtained MPM values and the measure proposed in this study are presented in Table **10**.

Table 10. The obtained MPM values for both methods.

Method	MPM proposed by Aladag&Turksen (2015)	The proposed performance measure
Aladag *et al.* [3]	0.48376	406.5408
Egrioglu [17]	0.37031	216.9368
Proportions	1.31	1.87

According to Table **7** the method proposed by Egrioglu [17] gives more accurate forecasts than those obtained from Aladag *et al.* [3] in terms of MPM. It is clearly observed that the both performance measure pick the same method as the best one for the enrollment data. According to Table **10**, the fuzzy time series forecasting approach proposed by Egrioglu [17] produced more accurate results than those obtained from the method proposed by Aladag *et al.* [3]. In the last row of Table **10**, some proportions are given. For instance, the proportion of MPM value produced by Aladag *et al.* [3] to MPM value obtained from Egrioglu [17] is 1.31. The proportion value for the proposed method is 1.87. According to these proportion values, it can be said that the proposed performance measure is more sensitive than MPM.

Secondly, IMKB100 time series is forecasted by using both fuzzy time series approaches proposed by Aladag *et al.* [3] and Egrioglu [17]. Daily observations of the data are between 1/2/2013 and 31/05/2013 so it includes 106 observations. The last 7 observations are used as the test set while the rest of the data is being used for training. The best forecasts for Aladag *et al.* [3] and Egrioglu [17] are obtained for 7 and 5 clusters, respectively. Thus, membership values for 7 and 5 clusters are obtained by using fuzzy c-means method. For the observations in the test set, all membership values are depicted in Tables **11** and **12**.

Table 11. The membership values of observations in the test set for 7 clusters.

Date	Test Data	Cluster1	Cluster2	Cluster3	Cluster4	Cluster5	Cluster6	Cluster7
23/05/2013	91351	0.00180	0.00302	0.02994	0.00104	0.96311	0.00044	0.00065
24/05/2013	91016	0.00761	0.01317	0.18023	0.00429	0.79027	0.00177	0.00265
27/05/2013	90547	0.01215	0.02211	0.59963	0.00658	0.35296	0.00261	0.00397
28/05/2013	89916	0.00153	0.00304	0.97965	0.00078	0.01426	0.00029	0.00045
29/05/2013	87175	0.10854	0.69467	0.10775	0.03211	0.03466	0.00790	0.01436
30/05/2013	87170	0.10821	0.69678	0.10657	0.03195	0.03434	0.00786	0.01428
31/05/2013	85990	0.01908	0.97296	0.00247	0.00286	0.00107	0.00051	0.00103

Table 12. The membership values of observations in the test set for 5 clusters.

Date	Test Data	Cluster 1	Cluster 2	Cluster 3	Cluster 4	Cluster 5
23/05/2013	91351	0.00249	0.00373	0.01270	0.97489	0.00618
24/05/2013	91016	0.00069	0.00105	0.00376	0.99273	0.00177
27/05/2013	90547	0.00009	0.00014	0.00055	0.99897	0.00024
28/05/2013	89916	0.00370	0.00587	0.02617	0.95364	0.01061
29/05/2013	87175	0.02418	0.04489	0.64910	0.17265	0.10919

(Table 12) contd.....

Date	Test Data	Cluster 1	Cluster 2	Cluster 3	Cluster 4	Cluster 5
30/05/2013	87170	0.02412	0.04480	0.65047	0.17156	0.10906
31/05/2013	85990	0.00534	0.01105	0.93292	0.01621	0.03448

All membership values and the corresponding forecasts obtained from both methods are presented in Tables **13** and **14**.

Table 13. The membership values of predictions obtained from the method proposed by Aladag *et al.* [3].

Date	Aladag *et al.* [3]	Cluster1	Cluster2	Cluster3	Cluster4	Cluster5	Cluster6	Cluster7
23/05/2013	90674.29	0.01087	0.17345	0.06563	0.01613	0.06563	0.33415	0.33415
24/05/2013	91640.69	0.00901	0.17135	0.01307	0.00132	0.01307	0.37179	0.42039
27/05/2013	90674.29	0.00872	0.16582	0.07613	0.00556	0.07613	0.33382	0.33382
28/05/2013	89175.35	0.00789	0.13492	0.16109	0.00845	0.22921	0.22921	0.22921
29/05/2013	89175.35	0.00395	0.06480	0.11668	0.00084	0.27124	0.27124	0.27124
30/05/2013	85557.97	0.03797	0.03797	0.24302	0.21622	0.24302	0.03797	0.18384
31/05/2013	85557.23	0.03781	0.03781	0.24348	0.21597	0.24348	0.03781	0.18363

Table 14. The membership values of predictions obtained from the method proposed by Egrioglu [17].

Date	Egrioglu [17]	Cluster1	Cluster2	Cluster3	Cluster4	Cluster5
23/05/2013	90668.59	0.00160	0.00560	0.30006	0.00560	0.68715
24/05/2013	90668.59	0.00084	0.00301	0.34070	0.00301	0.65244
27/05/2013	90668.59	0.00010	0.00039	0.30368	0.00039	0.69544
28/05/2013	90668.59	0.00076	0.00271	0.30672	0.00271	0.68710
29/05/2013	86709.33	0.00041	0.00159	0.49820	0.00159	0.49820
30/05/2013	86089.64	0.01463	0.06522	0.42747	0.06522	0.42747
31/05/2013	85366.59	0.00555	0.32572	0.01730	0.32572	0.32572

Now it is possible to calculate the proposed performance measure based on the difference between the membership values of the observations and of the forecasts. In Table **15**, the obtained values of the criterion proposed in this research are summarized. Also, RMSE and MPM values obtained for both approaches are given in this table.

Table 15. The obtained values of measures for both methods.

Method	RMSE	MPM	The proposed measure
Aladag *et al.* [3]	1082	0.5361	30.0692
Egrioglu [17]	648	1.0409	94.3678
Proportion		1.9416	3.1384

According to Table **15**, the method suggested by Egrioglu [17] produced the best results in terms of RMSE which is a conventional criterion. On the other hand, the method suggested by Aladag *et al.* [3] produced the best results in terms of MPM and the proposed measure which are advanced criteria. As mentioned before, MPM and the suggested performance measure are based on the membership values in accordance with the nature of fuzzy logic. In terms of these advanced performance measures, it can be said that Aladag *et al.* [3] approach should be used to get accurate forecasts for IMKB100 time series. The both criteria is picked the same method. However, if the proportions is examined, it is clearly seen that the proposed one is more sensitive.

CONCLUDING REMARKS

There are many different fuzzy systems to solve real world problems from different fields. There is a common point for each of fuzzy systems applications. It is the evaluation of the fuzzy systems. One way to evaluate the performance of a fuzzy system is to use a criterion. In some cases, there are more than one model based on fuzzy logic. Then, all alternative models should be evaluated to pick the best model that gives the best performance. For instance, in fuzzy time series applications, it is tried to be determined the best model that produces the most accurate forecasts. In most of the studies available in the literature, conventional criteria such as MSE, RMSE or MAPE are utilized to measure the forecasting performance of fuzzy time series models. Such criteria are calculated using crisp values with ignoring the membership values. However, when models are evaluated, the membership values should be employed since these values carries important information. Ignoring the membership values can lead to misleading evaluation results. Therefore, an alternative performance criterion which is calculated using the membership values is proposed in this study. A new distance measure is firstly introduced then, a novel performance criterion using this measure is suggested to evaluate fuzzy systems.

Using a conventional performance measure calculated over defuzzified forecasts can bring some disadvantages such as an increase in error, with inconsistent results, and unnecessary operations. Because of all of these reasons, a performance measure in accordance with the nature of fuzzy logic should be used to evaluate fuzzy systems. On the other hand, there is only one performance measure available in the literature which takes membership values into consideration. It was proposed by Aladag and Turksen [2] and it is called MPM. Therefore, a new performance measure in accordance with the nature of fuzzy logic is proposed in this study to evaluate fuzzy logic based systems.

In this study, first of all, a new distance measure which is calculated over membership values is firstly defined. The suggested distance is computed by using membership values of targets and outputs of a fuzzy systems. Then, a new criterion is proposed by using this new distance measure. The algorithm of the proposed performance measure was coded in Matlab computer package. In order to show the applicability of the proposed performance measure, it is applied to two real world data in the implementation by using the coded program. Index 100 in stocks and bonds exchange market of İstanbul and the enrolment data of Alabama are utilized in the application. Two different fuzzy time series forecasting approaches proposed by Aladag *et al.* [3] and Egrioglu [17] are applied to these real world time series. And, all results obtained from both methods are evaluated by using the proposed performance measure. For the aim of comparison, MSE and RMSE as conventional criteria and MPM as an advanced measure are also employed to evaluate these fuzzy time series approaches. All obtained results were given and interpreted in the previous section. In the application, all computations are performed in Matlab computer package. As a result of the implementation, it is seen that the proposed performance measure is a good alternative criterion to evaluate fuzzy systems. In the application, we focused on fuzzy time series. In fuzzy time series approaches, targets and outputs of a system are observations and forecasts, respectively. On the other hand, the proposed criterion can be easily used for other fuzzy systems since it can be smoothly computed by using the membership values obtained from any fuzzy system studies.

There are many performance criteria available in the literature. It is not possible to say that one of them is the best. For example, no one can claim that RMSE is better MAPE or not. In this study, it is also not claimed that the proposed measure is better than other criteria available in the literature. Making such a comparison clearly depends on decision makers' further personal criteria of assessment. However, some problems can arise when conventional criteria such as MSE, RMSE or MAPE are used since it is necessary to employ a method for defuzzification and the membership values are ignored. To deal with these problems, in this study, we propose a new alternative performance measure which is computed based on the membership values. If decision makers would like to take advantage of using the proposed criterion, we suggest that they exploit it. In addition, the proposed measure can be used with other criteria in order to make sound decisions.

CONFLICT OF INTEREST

The authors (editor) declares no conflict of interest, financial or otherwise.

ACKNOWLEDGEMENTS

The first author wants to express his gratitude to the Scientific and Technological Research Council of Turkey (TUBİTAK) for financial support during his study visit (one year) to University of Toronto.

REFERENCES

[1]   C.H. Aladag, "Using multiplicative neuron model to establish fuzzy logic relationships", *Expert Syst. Appl.,* vol. 40, no. 3, pp. 850-853, 2013.
        [http://dx.doi.org/10.1016/j.eswa.2012.05.039]

[2]   C.H. Aladag, and I.B. Turksen, "A novel membership value based performance measure", *J. Intell. Fuzzy Syst.,* vol. 28, no. 2, pp. 919-928, 2015.

[3]   C.H. Aladag, E. Egrioglu, U. Yolcu, and A.Z. Dalar, "A new time invariant fuzzy time series forecasting method based on particle swarm optimization", *Appl. Soft Comput.,* vol. 12, no. 10, pp. 3291-3299, 2012.
        [http://dx.doi.org/10.1016/j.asoc.2012.05.002]

[4]   C.H. Aladag, E. Egrioglu, U. Yolcu, and V.R. Uslu, "A high order seasonal fuzzy time series model and application to international tourism demand of Turkey", *J. Intell. Fuzzy Syst.,* vol. 26, pp. 295-302, 2014.

[5]   C.H. Aladag, U. Yolcu, E. Egrioglu, and B. Eren, "Fuzzy lagged variable selection in fuzzy time series with genetic algorithms", *Appl. Soft Comput.,* vol. 22, pp. 465-473, 2014.
        [http://dx.doi.org/10.1016/j.asoc.2014.03.028]

[6]   J.S. Armstrong, and F. Collopy, "Error Measures for Generalizing about Forecasting Methods: Empirical Comparisons. Reprinted version with permission form", *Int. J. Forecast.,* vol. 8, pp. 69-80, 1992.
        [http://dx.doi.org/10.1016/0169-2070(92)90008-W]

[7]   M. Avazbeigi, S.H. Doulabi, and B. Karimi, "Choosing the appropriate order in fuzzy time series: A new N-factor fuzzy time series for prediction of the auto industry production", *Expert Syst. Appl.,* vol. 37, no. 8, pp. 5630-5639, 2010.
        [http://dx.doi.org/10.1016/j.eswa.2010.02.049]

[8]   U.C. Benz, P. Hofmann, G. Willhauck, I. Lingenfelder, and M. Heynen, "Multi-resolution, object-oriented fuzzy analysis of remote sensing data for GIS-ready information", *ISPRS J. Photogramm. Remote Sens.,* vol. 58, pp. 239-258, 2004.
        [http://dx.doi.org/10.1016/j.isprsjprs.2003.10.002]

[9]   E. Cables, M.S. García-Cascales, and M.T. Lamata, "The LTOPSIS: An alternative to TOPSIS decision-making approach for linguistic variables", *Expert Syst. Appl.,* vol. 39, pp. 2119-2126, 2012.
        [http://dx.doi.org/10.1016/j.eswa.2011.07.119]

[10]   Q. Cai, D. Zhang, B. Wu, and S.C. Leung, "A Novel Stock Forecasting Model based on Fuzzy Time Series and Genetic Algorithm", *Procedia Comput. Sci.,* vol. 18, pp. 1155-1162, 2013.
        [http://dx.doi.org/10.1016/j.procs.2013.05.281]

[11] TL Chen, CH Cheng, and HJ Teoh, "High-order fuzzy time-series based on multi-period adaptation model for forecasting stock markets", *Physica A: Statistical Mechanics and its Applications,* vol. 387, no. 4, pp. 876-888, 2008.
[http://dx.doi.org/10.1016/j.physa.2007.10.004]

[12] S.M. Chen, and C.D. Chen, *IEEE Trans. Fuzzy Syst.,* vol. 19, no. 1, pp. 1-12, 2011.
[http://dx.doi.org/10.1109/TFUZZ.2010.2073712]

[13] M.Y. Chen, and B.T. Chen, "A hybrid fuzzy time series model based on granular computing for stock price forecasting", *Inf. Sci.,* vol. 294, pp. 227-241, 2015.
[http://dx.doi.org/10.1016/j.ins.2014.09.038]

[14] S.M. Chen, and P.Y. Kao, "TAIEX forecasting based on fuzzy time series, particle swarm optimization techniques and support vector machines", *Inf. Sci.,* vol. 247, no. 20, pp. 62-71, 2013.
[http://dx.doi.org/10.1016/j.ins.2013.06.005]

[15] Y.C. Cheng, and S.T. Li, "Fuzzy time series forecasting with a probabilistic smoothing hidden Markov model", *IEEE Trans. Fuzzy Syst.,* vol. 20, no. 2, pp. 291-304, 2012.
[http://dx.doi.org/10.1109/TFUZZ.2011.2173583]

[16] E. Egrioglu, C.H. Aladag, M.A. Basaran, U. Yolcu, and V.R. Uslu, "A new approach based on the optimization of the length of intervals in fuzzy time series", *J. Intell. Fuzzy Syst.,* vol. 22, pp. 15-19, 2011.

[17] E. Egrioglu, "A New Time invariant fuzzy time series forecasting method based on genetic algorithm", *Advances in Fuzzy Systems,* 2012.
[PMID: 785709]

[18] A. Fattouh, and F. Fouz, "A Two-Stage Representation of Fuzzy Systems", *International Journal of Engineering Research and Applications,* vol. 2, no. 3, pp. 2660-2665, 2012.

[19] R.J. Hyndman, and A.B. Koehler, "Another look at measures of forecast accuracy", *Int. J. Forecast.,* vol. 22, pp. 679-688, 2006.
[http://dx.doi.org/10.1016/j.ijforecast.2006.03.001]

[20] C Kahraman, M Yavuz, and I Kaya, "Fuzzy and grey forecasting techniques and their applications in production systems Studies in Fuzziness and Soft Computing", *Production Engineering and Management under Fuzziness,* vol. 252, pp. 1-24, 2010.

[21] W.J. Lee, and J. Hong, "A hybrid dynamic and fuzzy time series model for mid-term power load forecasting", *Int. J. Electr. Power Energy Syst.,* vol. 64, pp. 1057-1062, 2015.
[http://dx.doi.org/10.1016/j.ijepes.2014.08.006]

[22] S.T. Li, Y.C. Cheng, and S.Y. Lin, "A FCM-based deterministic forecasting model for fuzzy time series", *Comput. Math. Appl.,* vol. 56, no. 1, pp. 3052-3063, 2008.
[http://dx.doi.org/10.1016/j.camwa.2008.07.033]

[23] W. Lu, X. Chen, W. Pedrycz, X. Liu, and J. Yang, "Using interval information granules to improve forecasting in fuzzy time series", *Int. J. Approx. Reason.,* vol. 57, pp. 1-18, 2015.
[http://dx.doi.org/10.1016/j.ijar.2014.11.002]

[24] Z. Ma, Q. Dai, and N. Liu, "Several novel evaluation measures for rank-based ensemble pruning with applications to time series prediction", *Expert Syst. Appl.,* vol. 42, no. 1, pp. 280-292, 2015.
[http://dx.doi.org/10.1016/j.eswa.2014.07.049]

[25] G. Nan, S. Zhou, J. Kou, and M. Li, "Heuristic bivariate forecasting model of multi-attribute fuzzy time series based on fuzzy clustering", *Int. J. Inf. Technol. Decis. Mak,* vol. 11, no. 1, pp. 167-195, 2012.
[http://dx.doi.org/10.1142/S0219622012500083]

[26] G. Panchal, A. Ganatra, Y.P. Kosta, and D. Panchal, "Searching most efficient neural network architecture using Akaike's information criterion (AIC)", *Int. J. Comput. Appl.,* vol. 1, no. 5, pp. 975-

8887, 2010.

[27]   W. Qiu, X. Liu, and L. Wang, "Forecasting shanghai composite index based on fuzzy time series and improved C-fuzzy decision trees", *Expert Syst. Appl.,* vol. 39, no. 9, pp. 7680-7689, 2012.
       [http://dx.doi.org/10.1016/j.eswa.2012.01.051]

[28]   MV Shcherbakov, A Brebels, and NL Shcherbakov, *Survey of Forecast Error Measures, World Applied Sciences Journal 24 (Information Technologies in Modern Industry, Education & Society),* pp. 171-176, 2013.

[29]   P. Singh, and B. Borah, "An efficient time series forecasting model based on fuzzy time series", *Eng. Appl. Artif. Intell.,* vol. 26, no. 10, pp. 2443-2457, 2013.
       [http://dx.doi.org/10.1016/j.engappai.2013.07.012]

[30]   Q. Song, and B.S. Chissom, "Fuzzy time series and its models", *Fuzzy Sets Syst.,* vol. 54, pp. 269-277, 1993.
       [http://dx.doi.org/10.1016/0165-0114(93)90372-O]

[31]   Q. Song, and B.S. Chissom, "Forecasting enrollments with fuzzy time series- Part I", *Fuzzy Sets Syst.,* vol. 54, pp. 1-10, 1993.
       [http://dx.doi.org/10.1016/0165-0114(93)90355-L]

[32]   Q. Song, and B.S. Chissom, "Forecasting enrollments with fuzzy time series - Part II", *Fuzzy Sets Syst.,* vol. 62, no. l, pp. 1-8, 1994.
       [http://dx.doi.org/10.1016/0165-0114(94)90067-1]

[33]   C.C. Wang, "A comparison study between fuzzy time series model and ARIMA model for forecasting Taiwan export", *Expert Syst. Appl.,* vol. 38, no. 8, pp. 9296-9304, 2011.
       [http://dx.doi.org/10.1016/j.eswa.2011.01.015]

[34]   U. Yolcu, C.H. Aladag, E. Egrioglu, and V.R. Uslu, "Time-series forecasting with a novel fuzzy time-series approach: an example for Istanbul stock market", *J. Stat. Comput. Simul.,* vol. 83, no. 4, pp. 597-610, 2013.
       [http://dx.doi.org/10.1080/00949655.2011.630000]

[35]   U. Yolcu, O. Cagcag, C.H. Aladag, and E. Egrioglu, "An enhanced fuzzy time series forecasting method based on artificial bee colony algorithm", *J. Intell. Fuzzy Syst.,* vol. 26, pp. 2627-2637, 2014.

[36]   T.H Yu, and K-H. Huarng, "A Neural Network-based Fuzzy Time Series Model to Improve Forecasting", *Expert Syst. Appl.,* vol. 37, pp. 3366-3372, 2010.
       [http://dx.doi.org/10.1016/j.eswa.2009.10.013]

[37]   L.A. Zadeh, "Fuzzy Sets", *Inf. Control,* vol. 8, no. 3, pp. 338-353, 1965.
       [http://dx.doi.org/10.1016/S0019-9958(65)90241-X]

[38]   L.A. Zadeh, "The concept of linguistic variable and its application to approximate reasoning", *Inf. Sci.,* vol. 8, pp. 199-249, 1975.
       [http://dx.doi.org/10.1016/0020-0255(75)90036-5]

CHAPTER 2

A New Fuzzy Time Series Forecasting Model with Neural Network Structure

Eren Bas[*] and Erol Egrioglu

Department of Statistics, Faculty of Arts and Sciences, Giresun University, Giresun, Turkey

Abstract: Non-probabilistic forecasting methods are one of the most popular forecasting methods in recent years. Fuzzy time series methods are non-probabilistic and non-linear methods. Although these methods have superior forecasting performance, linear autoregressive models have better forecasting performance than fuzzy time series methods for some real-life time series. In this paper, a new hybrid forecasting method that contains stochastic approach based on an autoregressive model and fuzzy time series forecasting model was proposed in a network structure. Fuzzy c means method is used in fuzzification stage of the proposed method and also the proposed method is trained by using particle swarm optimization. The proposed method is applied to a well-known real-life time series data and it is proved that the proposed method has best forecasting performance when compared with some other studies suggested in the literature.

Keywords: Autoregressive model, Forecasting, Fuzzy c-means, Fuzzy time series, Non-linear time series, Particle swarm optimization.

INTRODUCTION

In the literature, there are many forecasting methods used in almost all areas. Forecasting methods are known as non-probabilistic and probabilistic approaches. Probabilistic approaches are also known as classical time series approaches. Besides, autoregressive model, it is also a linear model, is a popular method among probabilistic methods. This model summarized in the study of Box and Jenkins [1]. Besides this model, fuzzy methods and artificial neural networks (ANN) are some of non-probabilistic approaches. Methods contains fuzzy logic are based on fuzzy set theory which was proposed by Zadeh [2]. In the literature, there are some methods for forecasting. These methods are fuzzy inference systems, fuzzy regression methods, fuzzy function approaches and fuzzy time series methods.

---

[*] **Corresponding author Eren Bas:** Giresun University, Faculty of Arts and Science, Department of Statistics, 28200, Giresun, Turkey; E-mail: eren.bas@giresun.edu.tr

Cagdas Hakan Aladag (Ed.)

The studies for fuzzy inference systems were proposed by Mamdani and Assilian [3], Takagi and Sugeno [4] and Jang [5]. Fuzzy inference systems are based on rules. But, they have an important problem. This problem is determination of the rules. To get over this problem, fuzzy function approach was proposed by Turksen [6].

Fuzzy time series methods become one of the most popular methods for forecasting of time series in recent years. Song and Chissom [7 - 9] mentioned the basis of fuzzy time series models. Fuzzy time series methods have three stages; fuzzification, determination of fuzzy relations and defuzzification. There are many methods for each of these stages to improve the forecasting performance.

In the literature, the decomposition of universe of discourse was usually used in the fuzzification stage and the intervals of it were determined arbitrarily as in the studies of [7 - 10]. In addition [11, 12], emphasized the significance of the interval length on the forecasting performance and suggested two methods based on the mean and the distribution to find intervals and also varied optimization techniques have been used in other studies in the literature. And also [13, 14], suggested forming the problem of finding intervals as an optimization problem. Artificial intelligence optimization algorithms are tools used in fuzzification stage. Genetic algorithm was used to find the interval lengths in the studies of [15, 16] and also particle swarm optimization (PSO) was used in the studies of [17 - 23]. In addition to these studies, differential evolution algorithm was used to find the interval lengths in the studies of [24].

In the stage of determination of fuzzy relations, generally, researchers used matrix operations proposed by [7 - 9]. And also, fuzzy logic relations group table was used in some other studies and the study of [10]. Besides these studies, ANN was used in recent years in the studies of [11, 25 - 29] to determine the fuzzy relations. The other studies in this stage were proposed by [26, 30 - 32].

In defuzzification stage, the centroid method is the most commonly method used in the literature. And also, adaptive expectation method was used in recent years [10, 12, 31, 33].

For some time series, linear models have outstanding results when linear part of time series is better than nonlinear part. Besides, the contrary of this statement is also true. However, in either case, one of these parts is not considered important. In this way, it may cause fallacious results. To overcome this like problem, there are many hybrid methods such as proposed by Tseng and coworkers [34], Zhang and coworkers [35], BuHamra and coworkers [36], Pai and Lin [37], Wang and coworkers [38], Jain and Kumar [39], Chen and Wang [40], Aladag and coworkers [41] and Lee and Tong [42]. These hybrid methods have two parts.

One of them is linear model and the other one is ANN method. Besides these studies, Yolcu and coworkers [43] combined autoregressive and MLP-ANN in a network structure.

In this paper, a hybrid forecasting method using autoregressive model and fuzzy time series forecasting model in a same network structure is proposed. Then, the proposed method is analyzed with a real world time series data and the obtained results are compared with other results obtained from some other studies in the literature. The other parts of the paper are the proposed method, application and conclusions and discussions parts.

PROPOSED METHOD

In recent years, there are various studies in which fuzzy time series approaches are utilized for forecasting of fuzzy time series. Many studies in the literature have been proved that fuzzy time series forecasting models give superior forecasting performance for the forecasting of linear or non-linear real life time series. On the other hand, autoregressive models produce better forecasting results for real world time series that have a linear structure. Since these approaches have pearls and pitfalls, a hybrid approach of these methods can give more successful results for the analysis of both linear and nonlinear time series. For this reason, a novel hybrid forecasting method was suggested in this paper. The proposed method is network-structured and it is a blend of fuzzy time series, autoregressive and combination part. This proposed method was proposed in Bas and coworkers [44].

Algorithm 1. Computation algorithm for the output of FTS-N

Step 1. Computation of Fuzzy Time Series Part (FTSP)

In Step 1, output of fuzzy time series is calculated. The phase is applied as follows;

Step 1.1. Fuzzify observations of time series by using Fuzzy Clustering Method (FCM).

In the first layer of FTSP, FCM clustering algorithm is applied to time series. Let c be the number of fuzzy sets, such that $2 \leq c \leq n$ where n is the number of observations. FCM clustering algorithm (c is the number of fuzzy sets) is applied to the time series including crisp values. The center value for each fuzzy set is calculated. Then, the degrees for each observation, which denote a degree of belonging to a fuzzy set for that observation, are computed according to the obtained center values of fuzzy sets. Finally, ordered fuzzy sets L_r ($r = 1,2,\ldots,c$) are obtained according to the ascending ordered centers, which are denoted by v_r ($r =$

1,2,…,c).

Bezdek [45] firstly introduced FCM clustering. It is the most preferred clustering algorithm. In this clustering algorithm, fuzzy clustering is performed by minimizing the least squared errors within groups. Let u_{ij} be the membership values, v_i be the center of cluster, n be the number of variables, and c be the number of clusters. Then the objective function, which is tried to be minimized in fuzzy clustering, is

$$J_\beta(X,V,U) = \sum_{i=1}^{c}\sum_{j=1}^{n} u_{ij}^\beta d^2(x_j,v_i) \qquad (1)$$

where, X, V and U are data matrix, the center of clusters matrix and the matrix of membership values, respectively, β is a constant ($\beta > 1$) and called the fuzzy index. $d(x_j,v_i)$ is a similarity measure between an observation and the center of corresponding fuzzy cluster. The objective function J_β is minimized subject to constraints given below.

$$0 \le u_{ij} \le 1 \qquad ,\forall i,j$$

$$0 < \sum_{j=1}^{n} u_{ij} \le n \qquad ,\forall i \qquad (2)$$

$$\sum_{i=1}^{c} u_{ij} = 1 \qquad ,\forall j$$

Step 1.2. Fuzzy lagged variables are obtained from fuzzy time series.

Lagged variables are obtained from the outputs of the previous layer in the second layer of FTSP and inputs of the next layer are computed. In other words, lagged variables are calculated by using fuzzy time series *(F(t))* created from FCM method.

Step 1.3. Aggregation of fuzzy inputs.

In the third layer of FTSP, a membership value corresponding to each fuzzy set is calculated by using the intersection operator to membership values of fuzzy lagged variables. The output of the third layer is a vector represented by $\mu(t)$ including c elements for t^{th} learning sample.

Step 1.4. Calculation of FTSP output

A multiplicative neuron model is utilized in this step. Therefore, the inputs produced by the previous layer are aggregated by means of product function. The output value for FTSP is calculated given below.

$$output_t^{FTSP} = \prod_{i=1}^{c}(w_i\mu_i^t + b_i) \quad t = 1,2,\dots,n \tag{3}$$

Step 2. Computation of Autoregressive Part (ARP)

In this part, the values corresponding to the variables in the lagged input layer are aggregated by means of additive functions. Then, the output value for ARP can be computed by using the formula given below.

$$output_t^{ARP} = \sum_{j=1}^{p} v_j x_{t-j} + v_b \,,t = 1,2,\dots,n \tag{4}$$

Step 3. Computation of network part

Outputs of FTSP and ARP compose the inputs of combination part. In this part, inputs from FTSP and ARP are aggregated by means of additive function.

$$output_t^{FTS-N} = \hat{X}_t = r_1 output_t^{FTSP} + r_2 output_t^{ARP} + r_b \quad t = 1,2,\dots,n \tag{5}$$

When the network part is being trained, PSO is employed. PSO proposed by Kennedy and Eberhart [46] is an artificial intelligence optimization technique. In the literature, there are several studies using different types of PSO for the training of ANN. The PSO algorithm used to train the network part is a modified algorithm. This algorithm for training is given in Algorithm 2.

Algorithm 2. PSO algorithm used to train of the proposed method

Step 1. The parameters of PSO are specified.

First of all, the parameters of the PSO algorithm are determined. These parameters direct PSO algorithm are pn, vm, c_1, c_2 and w. Let c_1 and c_2 represents cognitive and social coefficients, respectively, and w is the inertia parameter.

Step 2. Initial positions of each mth ($m = 1,2, \ldots, pn$) particles' positions and velocities are randomly determined and saved in vectors X_m and V_m shown below.

$$X_m = \{x_{m,1}, x_{m,2}, \cdots, x_{m,d}\}, m = 1,2, \cdots, pn \qquad (6)$$

$$V_m = \{v_{m,1}, v_{m,2}, \cdots, v_{m,d}\}, m = 1,2, \cdots, pn \qquad (7)$$

where j^{th} position of mth particle are represented by $x_{m,j}$ ($j=1,2,\ldots,d$). pn and d represents the number of particles in a swarm and positions, respectively.

The initial positions and velocities of each particle in a swarm are randomly obtained by utilizing uniform distribution $(0,1)$ and $(-vm, vm)$, respectively.

The positions of a particle in the swarm are the weights in the network part. Positions of the particles are shown in Fig. (**1**).

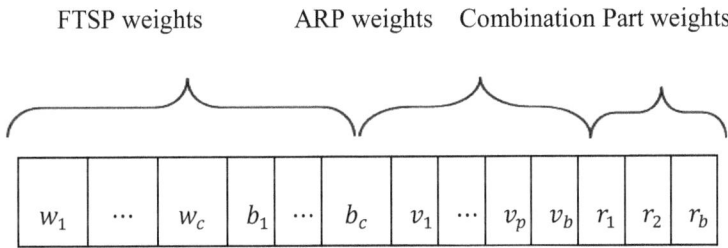

Fig. (1). Structure of a particle.

Step 3. Fitness function values are calculated. Fitness function values for each particle are calculated. Mean Square Error (MSE) given in below is used as evaluation function.

$$MSE = \frac{1}{n}\sum_{t=1}^{n}(X_t - \widehat{X_t})^2 \qquad (8)$$

where n represents the number of learning sample. The output value $(\widehat{X_t})$ of the proposed model is calculated by algorithm 1. X_t is the target value for the learning sample.

Step 4. $Pbest_m$ ($m = 1,2, \ldots, pn$) and *Gbest* are calculated based on evaluation function values obtained in the previous step. $Pbest_m$ is a vector stores the positions corresponding to the mth particle's best individual performance, and *Gbest* is the best particle, which has the best evaluation function value, found so far.

$$Pbest_m = (p_{m,1}, p_{m,2}, \cdots p_{m,d}), m = 1, 2, \cdots, pn \tag{9}$$

$$Gbest = \left(p_{g,1}, p_{g,2}, \cdots p_{g,d}\right) \tag{10}$$

Step 5. New values of positions and velocities are calculated. New values of positions and velocities for each particle are computed by using the formulas given in (11) and (12), respectively. If maximum iteration number is reached, the algorithm goes to *Step 3*; otherwise, it goes to *Step 6*.

$$v_{m,j}^{t+1} = \left[w \times v_{m,j}^t + c_1 \times rand_1 \times \left(p_{m,j} - x_{m,j}^t\right) + c_2 \times rand_2 \times \left(p_{g,j} - x_{m,j}^t\right)\right] \tag{11}$$

$$x_{m,j}^{t+1} = x_{m,j}^t + v_{m,j}^{t+1} \tag{12}$$

where $= 1, 2, \dots, pn, j = 1, 2, \dots, d$.

Step 6. The best solution is determined. The elements of *Gbest* are taken as the best weight values of the network part.

APPLICATION

For the evaluation of the proposed method, the time series data of beer consumption in Australia data observed between the years 1956 and 1994 whose graph is shown in Fig. (2) was analyzed. The time series data was analyzed with seasonal autoregressive integrated moving average (SARIMA), Winter's multiplicative exponential smoothing (WMES), Yolcu and coworkers [43], multiplicative neuron model based fuzzy time series method proposed by Aladag [47], fuzzy functions approach proposed by Turksen [6] and the proposed method.

The last sixteen observation of time series with 148 observations were used as test data and the rest of the observations were used as training data.

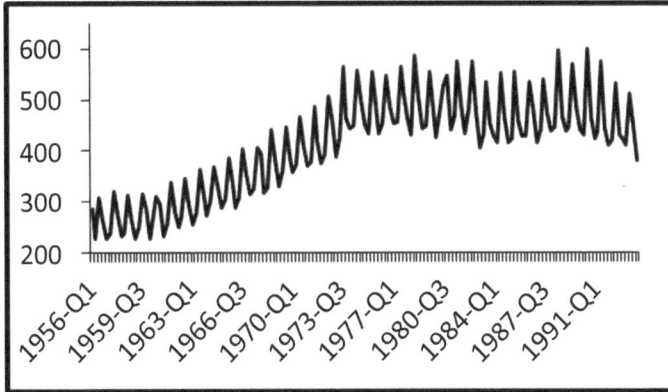

Fig. (2). Time Series of Beer Consumption in Australia.

At the end of all analysis for the methods given above, Root Mean Square Error (RMSE) and Mean Absolute Percentage Error (MAPE) criteria calculated for the test set were employed. RMSE and MAPE are calculated from the formulas (13) and (14), respectively.

$$RMSE = \sqrt{\frac{1}{n}\sum_{t=1}^{n}(X_t - \widehat{X_t})^2} \tag{13}$$

$$MAPE = \frac{1}{n}\sum_{t=1}^{n}\left|\frac{X_t - \widehat{X_t}}{X_t}\right| \tag{14}$$

When the proposed method is applied, model order (p) is shifted between 4 and 8; the number of fuzzy sets is changed between 5 and 15. Best-case results were obtained. In the process of PSO, $w = 0.9$, $c_1 = c_2 = 2$, $pn = 30$, $vm = 0.1$ and the maximum number of iterations is taken as 100.

Table **1** summarizes the results obtained from test set for the time series where the length of test set *(ntest)* is 16. In Table **1**, the best result of the proposed method was obtained from seventh-order model when the number of fuzzy sets is eleven. As seen in Table **1**, the proposed method gives the best result with respect to the forecasting performance.

Table 1. Comparison of the results for test set.

Test Data	SARIMA	WMES	[43]	[47]	[6]	[44]
430.50	452.72	453.91	449.92	437.50	446.20	423.29
600.00	578.29	575.22	574.28	537.50	580.12	557.93

(Table 1) contd.....

Test Data	SARIMA	WMES	[43]	[47]	[6]	[44]
464.50	487.70	502.32	481.47	437.50	483.04	478.08
423.60	446.28	444.73	442.79	437.50	442.97	425.13
437.00	456.77	459.66	445.12	437.50	444.74	441.29
574.00	583.51	582.48	571.97	537.50	579.90	566.27
443.00	492.13	508.64	472.76	487.50	468.01	455.71
410.00	450.36	450.31	416.36	437.50	418.98	422.05
420.00	461.01	465.4	428.63	437.50	431.60	419.61
532.00	588.96	589.74	559.89	562.50	559.41	545.19
432.00	496.77	514.96	445.75	462.50	444.08	440.34
420.00	454.64	455.89	390.25	412.50	394.99	398.00
411.00	465.46	471.15	412.38	437.50	409.72	392.10
512.00	594.71	597.00	533.19	537.50	525.60	507.73
449.00	501.67	521.28	442.13	437.50	438.91	435.48
382.00	459.17	461.46	405.08	412.50	409.07	405.22
RMSE	47.0367	53.3295	18.7888	29.1381	17.3926	16.2377
MAPE	0.0949	0.1072	0.0357	0.0532	0.0345	0.0277

CONCLUSIONS AND DISCUSSIONS

In this study, a hybrid forecasting method which combines an autoregressive model and fuzzy time series forecasting model in a network structure was proposed. To evaluate the performance of the proposed method, the proposed method was applied to a real-world time series data. It is understood as a result of the application of the proposed method that the proposed method has better forecasting performance when compared with the some other forecasting methods proposed in the literature.

In addition, the method proposed in this paper is an effective hybrid forecasting approach combines fuzzy time series and autoregressive method. By introducing this hybrid approach, a novel network method that can model both nonlinear and linear structures was suggested. The main objective of the future studies is obtaining confidence intervals for forecasts and having input choice made by stochastic method rather than trial and error method.

CONFLICT OF INTEREST

The authors (editor) declares no conflict of interest, financial or otherwise.

ACKNOWLEDGEMENTS

Declared none.

REFERENCES

[1] G.E. Box, and G.M. Jenkins, *Time Series Analysis: Forecasting and Control.* Holdan-Day: San Francisco, CA, 1976.

[2] L.A. Zadeh, "Fuzzy sets", *Inf. Control,* vol. 8, pp. 338-353, 1965.
[http://dx.doi.org/10.1016/S0019-9958(65)90241-X]

[3] E.H. Mamadani, and S. Assilian, "An experiment in linguistic synthesis with a fuzzy logic controller", *Int J Man Math Stat,* vol. 7, no. 1, pp. 1-13, 1975.

[4] T. Takagi, and M. Sugeno, "Fuzzy identification of systems and its applications to modeling and control", *IEEE Trans. Syst. Man Cybern.,* vol. 15, no. 1, pp. 116-132, 1985.
[http://dx.doi.org/10.1109/TSMC.1985.6313399]

[5] J.S. Jang, "ANFIS: Adaptive network based fuzzy inference system", *IEEE Trans. Syst. Man Cybern.,* vol. 23, no. 3, pp. 665-685, 1993.
[http://dx.doi.org/10.1109/21.256541]

[6] B. Turksen, "Fuzzy function with LSE", *Appl. Soft Comput.,* vol. 8, pp. 1178-1188, 2008.
[http://dx.doi.org/10.1016/j.asoc.2007.12.004]

[7] Q. Song, and B.S. Chissom, "Fuzzy time series and its models", *Fuzzy Sets Syst.,* vol. 54, pp. 269-277, 1993.
[http://dx.doi.org/10.1016/0165-0114(93)90372-O]

[8] Q. Song, and B.S. Chissom, "Forecasting enrollments with fuzzy time series - Part I", *Fuzzy Sets Syst.,* vol. 54, pp. 1-10, 1993.
[http://dx.doi.org/10.1016/0165-0114(93)90355-L]

[9] Q. Song, and B.S. Chissom, "Forecasting enrollments with fuzzy time series - Part II", *Fuzzy Sets Syst.,* vol. 62, no. 1, pp. 1-8, 1994.
[http://dx.doi.org/10.1016/0165-0114(94)90067-1]

[10] S.M. Chen, "Forecasting enrollments based on fuzzy time-series", *Fuzzy Sets Syst.,* vol. 81, pp. 311-319, 1996.
[http://dx.doi.org/10.1016/0165-0114(95)00220-0]

[11] K. Huarng, "Effective length of intervals to improve forecasting in fuzzy time-series", *Fuzzy Sets Syst.,* vol. 123, pp. 387-394, 2001.
[http://dx.doi.org/10.1016/S0165-0114(00)00057-9]

[12] K. Huarng, "Heuristic models of fuzzy time series for forecasting", *Fuzzy Sets Syst.,* vol. 123, pp. 369-386, 2001.
[http://dx.doi.org/10.1016/S0165-0114(00)00093-2]

[13] E. Egrioglu, C.H. Aladag, U. Yolcu, V.R. Uslu, and M.A. Basaran, "Finding an optimal interval length in high order fuzzy time series", *Expert Syst. Appl.,* vol. 37, pp. 5052-5055, 2010.
[http://dx.doi.org/10.1016/j.eswa.2009.12.006]

[14] E. Egrioglu, C.H. Aladag, M.A. Basaran, V.R. Uslu, and U. Yolcu, "A new approach based on the optimization of the length of intervals in fuzzy time series", *J. Intell. Fuzzy Syst.,* vol. 22, pp. 15-19, 2011.

[15] S.M. Chen, and N.Y. Chung, "Forecasting enrolments using high order fuzzy time series and genetic algorithms", *Int. J. Intell. Syst.,* vol. 21, pp. 485-501, 2006.
[http://dx.doi.org/10.1002/int.20145]

[16]   L.W. Lee, L.H. Wang, S.M. Chen, and Y.H. Leu, "Handling forecasting problems based on two factor high-order fuzzy time series", *IEEE Trans. Fuzzy Syst.,* vol. 14, no. 3, pp. 468-477, 2006.
[http://dx.doi.org/10.1109/TFUZZ.2006.876367]

[17]   F.P. Fu, K. Chi, W.G. Che, and Q.J. Zhao, "High-order difference heuristic model of fuzzy time series based on particle swarm optimization and information entropy for stock markets", *International Conference on Computer Design and Applications,* 2010.
[http://dx.doi.org/10.1109/ICCDA.2010.5541222]

[18]   Y.L. Huang, S.J. Horng, T.W. Kao, R.S. Run, J.L. Lai, R.J. Chen, I.H. Kuo, and M.K. Khan, "An improved forecasting model based on the weighted fuzzy relationship matrix combined with a PSO adaptation for enrollments", *Int. J. Innov. Comput., Inf. Control,* vol. 7, no. 7, pp. 4027-4046, 2011.

[19]   I-H. Kuo, S-J. Horng, T-W. Kao, T-L. Lin, C-L. Lee, and Y. Pan, "An improved method for forecasting enrollments based on fuzzy time series and particle swarm optimization", *Expert Syst. Appl.,* vol. 36, pp. 6108-6117, 2009.
[http://dx.doi.org/10.1016/j.eswa.2008.07.043]

[20]   I-H. Kuo, S-J. Horng, Y-H. Chen, R-S. Run, T-W. Kao, R-J. Chen, J-L. Lai, and T-L. Lin, "Forecasting TAIFEX based on fuzzy time series and particle swarm optimization", *Expert Syst. Appl.,* vol. 37, pp. 1494-1502, 2010.
[http://dx.doi.org/10.1016/j.eswa.2009.06.102]

[21]   S. Davari, M.H. Zarandi, and I.B. Turksen, "An Improved fuzzy time series forecasting model based on particle swarm internalization", *The 28th North American Fuzzy Information Processing Society Annual Conferences (NAFIPS 2009),* 2009 Cincinnati, Ohio, USA.
[http://dx.doi.org/10.1109/NAFIPS.2009.5156420]

[22]   J-I. Park, D-J. Lee, C-K. Song, and M-G. Chun, "TAIFEX and KOSPI 200 forecasting based on two factors high order fuzzy time series and particle swarm optimization", *Expert Syst. Appl.,* vol. 37, pp. 959-967, 2010.
[http://dx.doi.org/10.1016/j.eswa.2009.05.081]

[23]   L-Y. Hsu, S-J. Horng, T-W. Kao, Y-H. Chen, R-S. Run, R-J. Chen, J-L. Lai, and I-H. Kuo, "Temperature prediction and TAIFEX forecasting based on fuzzy relationships and MTPSO techniques", *Expert Syst. Appl.,* vol. 37, pp. 2756-2770, 2010.
[http://dx.doi.org/10.1016/j.eswa.2009.09.015]

[24]   V.R. Uslu, E. Bas, U. Yolcu, and E. Egrioglu, "A new fuzzy time series analysis approach by using differential evolution algorithm and chronologically determined weights", *Journal of Social and Economic Statistics,* vol. 2, no. 1, pp. 18-30, 2013.

[25]   K. Huarng, and H.K. Yu, "The application of neural networks to forecast fuzzy time series", *Physica A,* vol. 363, pp. 481-491, 2006.
[http://dx.doi.org/10.1016/j.physa.2005.08.014]

[26]   C.H. Aladag, M.A. Basaran, E. Egrioglu, U. Yolcu, and V.R. Uslu, "Forecasting in high order fuzzy time series by using neural networks to define fuzzy relations", *Expert Syst. Appl.,* vol. 36, pp. 4228-4231, 2009.
[http://dx.doi.org/10.1016/j.eswa.2008.04.001]

[27]   E. Egrioglu, C.H. Aladag, U. Yolcu, V.R. Uslu, and M.A. Basaran, "A new approach based on artificial neural networks for high order multivariate fuzzy time series", *Expert Syst. Appl.,* vol. 36, pp. 10589-10594, 2009.
[http://dx.doi.org/10.1016/j.eswa.2009.02.057]

[28]   E. Egrioglu, C.H. Aladag, U. Yolcu, M.A. Basaran, and V.R. Uslu, "A new hybrid approach based on SARIMA and partial high order bivariate fuzzy time series forecasting model", *Expert Syst. Appl.,* vol. 36, pp. 7424-7434, 2009.
[http://dx.doi.org/10.1016/j.eswa.2008.09.040]

[29]   E. Egrioglu, V.R. Uslu, U. Yolcu, M.A. Basaran, and C.H. Aladag, "A new approach based on artificial neural networks for high order bivariate fuzzy time series", *Applications of Soft Computing,* pp. 265-273, 2009c.

[30]   J. Sullivan, and W.H. Woodall, "A comparison of fuzzy forecasting and Markov modeling", *Fuzzy Sets Syst.,* vol. 64, no. 3, pp. 279-293, 1994.
[http://dx.doi.org/10.1016/0165-0114(94)90152-X]

[31]   C.H. Aladag, U. Yolcu, and E. Egrioglu, "A high order fuzzy time series forecasting model based on adaptive expectation and artificial neural networks", *Math. Comput. Simul.,* vol. 81, pp. 875-882, 2010.
[http://dx.doi.org/10.1016/j.matcom.2010.09.011]

[32]   U. Yolcu, C.H. Aladag, E. Egrioglu, and V.R. Uslu, "Time series forecasting with a novel fuzzy time series approach: an example for İstanbul stock market", *Journal of Computational and Statistics Simulation,* vol. 83, pp. 597-610, 2013.
[http://dx.doi.org/10.1080/00949655.2011.630000]

[33]   C.H. Cheng, T.L. Chen, H.J. Teoh, and C.H. Chiang, "Fuzzy time series based on adaptive expectation model for TAIEX forecasting", *Expert Syst. Appl.,* vol. 34, pp. 1126-1132, 2008.
[http://dx.doi.org/10.1016/j.eswa.2006.12.021]

[34]   F.M. Tseng, H.C. Yu, and G.H. Tzeng, "Combining neural network model with seasonal time series ARIMA model", *Technol. Forecast. Soc. Change,* vol. 69, pp. 71-87, 2002.
[http://dx.doi.org/10.1016/S0040-1625(00)00113-X]

[35]   G. Zhang, "Time series forecasting using a hybrid ARIMA and neural network model", *Neurocomputing,* vol. 50, pp. 159-175, 2003.
[http://dx.doi.org/10.1016/S0925-2312(01)00702-0]

[36]   "S.; Smaoui, N.; Gabr, M. The Box-Jenkins analysis and neural networks: prediction and time series modeling", *Appl. Math. Model.,* vol. 27, pp. 805-815, 2003.
[http://dx.doi.org/10.1016/S0307-904X(03)00079-9]

[37]   P.F. Pai, and C.S. Lin, "A hybrid ARIMA and support vector machines model in stock price forecasting", *Omega,* vol. 33, no. 6, pp. 497-505, 2005.
[http://dx.doi.org/10.1016/j.omega.2004.07.024]

[38]   W. Wang, P.H. Van Gelder, J.K. Vrijling, and J. Ma, "Forecasting daily stream flow using hybrid ANN models", *J. Hydrol. (Amst.),* vol. 324, pp. 383-399, 2006.
[http://dx.doi.org/10.1016/j.jhydrol.2005.09.032]

[39]   A. Jain, and A.M. Kumar, "Hybrid neural network models for hydrological time series forecasting", *Appl. Soft Comput.,* vol. 7, pp. 585-592, 2007.
[http://dx.doi.org/10.1016/j.asoc.2006.03.002]

[40]   K.Y. Chen, and C.H. Wang, "A hybrid SARIMA and support vector machines in forecasting the production values of the machinery industry in Taiwan", *Expert Syst. Appl.,* vol. 32, no. 1, pp. 254-264, 2007.
[http://dx.doi.org/10.1016/j.eswa.2005.11.027]

[41]   C.H. Aladag, E. Egrioglu, and C. Kadilar, "Forecasting nonlinear time series with a hybrid methodology", *Appl. Math. Lett.,* vol. 22, pp. 1467-1470, 2009.
[http://dx.doi.org/10.1016/j.aml.2009.02.006]

[42]   Y-S. Lee, and L-I. Tong, "Forecasting time series using a methodology based on autoregressive integrated moving average and genetic programming", *Knowl. Base. Syst.,* vol. 24, pp. 66-72, 2011.
[http://dx.doi.org/10.1016/j.knosys.2010.07.006]

[43]   U. Yolcu, C.H. Aladag, and E. Egrioglu, "A new linear & nonlinear artificial neural network model for time series forecasting", *Decision Support System Journals,* vol. 54, pp. 1340-1347, 2013.
[http://dx.doi.org/10.1016/j.dss.2012.12.006]

[44]   E. Bas, E. Egrioglu, C.H. Aladag, and U. Yolcu, "Fuzzy-time-series network used to forecast linear and nonlinear time series", *Appl. Intell.,* vol. 43, pp. 343-355, 2015.
[http://dx.doi.org/10.1007/s10489-015-0647-0]

[45]   J.C. Bezdek, *Pattern Recognition with Fuzzy Objective Function Algorithms.* Plenum Press: NY, 1981.
[http://dx.doi.org/10.1007/978-1-4757-0450-1]

[46]   J. Kennedy, and R. Eberhart, "Particle swarm optimization", In: *Proceedings of IEEE International Conference on Neural Networks* IEEE Press: Piscataway, NJ, USA, 1995.
[http://dx.doi.org/10.1109/ICNN.1995.488968]

[47]   C.H. Aladag, "Using multiplicative neuron model to establish fuzzy logic relationships", *Expert Syst. Appl.,* vol. 40, no. 3, pp. 850-853, 2013.
[http://dx.doi.org/10.1016/j.eswa.2012.05.039]

Advances in Time Series Forecasting, 2017, Vol. 2, 37-75 37

Two Factors High Order Non Singleton Type-1 and Interval Type-2 Fuzzy Systems for Forecasting Time Series with Genetic Algorithm

M.H. Fazel Zarandi[1,*], **M. Yalinezhaad**[1] and **I.B. Turksen**[2]

[1] *Department of Industrial Engineering and Management Systems, Amirkabir University of Technology (Polytechnic of Tehran), Tehran, Iran*

[2] *Department of Industrial Engineering, TOBB University of Economics and Technology, Sogutozu, Ankara, Turkey*

Abstract: This paper presents an efficient and simplified type-1 and interval type-2 non singleton fuzzy logic systems (NSFLSs) in order to obviate time series forecasting problems. These methods have applied non singleton fuzzification by Sharp Gaussian membership function, logical inference with the First-Infer-Then-Aggregate (FITA) approach and parametric defuzzification. Rules are generated based on high order fuzzy time series. In interval type-2 FLS, which can better handle uncertainties, type-2 sets are generated, using fuzzy normal forms by applying Yager Parametric classes of operators. Moreover, in these systems, some elements such as membership functions, operators and length of intervals affect the forecasting results. In addition, a method for tuning parameters of fuzzy logic systems with genetic algorithm is presented. Finally, the proposed methods are applied to predict the temperature and the Taiwan Stock Exchange (TAIEX). The results show the higher degree of accuracy of the model compared to the previous methods.

Keywords: Forecasting, Fuzzy Time Series, Genetic Algorithm, Interval Type-2.

INTRODUCTION

Intelligent systems have extensive applications in different areas. Especially, fuzzy logic system that was introduced by Zadeh [28] has equipped the intelligent methodologies with more desirable capabilities. Fuzzy logic systems (FLSs) are flexible and have the ability to model any nonlinear functions to some degrees of accuracy. Other techniques, such as neural networks or evolutionary algorithms can be easily used as complementary to fuzzy logic techniques.

* **Corresponding author M.H. Fazel Zarandi:** Department of Industrial Engineering and Management Systems, Amirkabir University of Technology (Polytechnic of Tehran), P.O. Box 15875-4413, Tehran, Iran; E-mail: zarandi@aut.ac.ir

Cagdas Hakan Aladag (Ed.)

A good area of application of fuzzy technology is Fuzzy Time Series (FTSs) that has been applied successfully for forecasting problems and can be used to deal with forecasting problem where historical data has linguistic values.

Li et al [12] proposed a method to forecast the weather, by applying fuzzy grade statistical theory. Later, Song and Chissom [19] proposed the concept of fuzzy time series based on the fuzzy set theory. They presented two fuzzy time series models: time invariant [19] and the time variant [20] fuzzy time series models to solve forecasting problems in enrolment of the University of Alabama and they achieved good forecasting results. In recent years, fuzzy time series methods have been implemented in numerous areas, such as forecasting stock price, university enrolments, temperature, *etc.* In order to make predictions, some researchers have used different complementary methods, like neural networks, clustering techniques, heuristic and stochastic models, *etc* [1, 2, 6, 24]. For instance, Lee *et al.* [9, 10] suggested methods for forecasting the temperature and the TAIFEX based on two factors high order fuzzy time series or, Wang and Chen [24] developed a method for temperature prediction and TAIFEX forecasting based on automatic clustering techniques and two factors high order fuzzy time series.

It should be noted that most of the existing fuzzy forecasting methods based on fuzzy time series do not use logical inference, but they use crisp inference, *i.e.*, they consider the midpoint of fuzzy set of consequent in each rule as the results. In real applications, crisp systems with crisp inference have some difficulties. The first drawback of these systems is that they use crisp inputs without any logical inference, which causes disability in handling behaviour of historical data in order to forecast future data. The second drawback of crisp models is that the rules have no parameters to be tuned, thus they are not trainable and cannot forecast data with good accuracy rate. The third drawback of crisp models is that if the rules, which are generated from future data, are corresponding to the rules of historical data, they can forecast them, otherwise they are not capable of forecasting. Therefore, with these drawbacks, the forecasting accuracy rates of crisp systems are not good enough.

On the other hand, type-2 FLSs, in which antecedent or consequent membership functions or both are type-2 fuzzy sets, can better handle uncertainties of data. However, most forecasting methods are usually based on type-1 fuzzy logic sets that the membership functions are crisp and are not capable to directly handle uncertainties.

In the literature there are limited works focusing on type-2 fuzzy time series. Huarng and Yu [6] presented a type-2 fuzzy time series model as an extension of a type-1 model for forecasting the TAIEX in which they attempted to use the highest and the lowest price of the TAIEX to assist forecasting the closing price. This type-2 model utilizes the fuzzy relationships established by a type-1 model based on type1observations. Liang and Mendel [13] presented an interval type-2 fuzzy time series for forecasting Mackey–Glass Chaotic Time Series. They assumed that the sampled time series is corrupted by uniformly distributed additive noise. They used four-order fuzzy time series rules for forecasting and used only two fuzzy sets for each antecedent, so the numbers of rules were 16. They compared the performance of the five forecasting FLSs then, they showed that the type-2NSFLS performs better.

In this research, we use NSFLS and following processes to develop and to overcome the drawbacks of crisp fuzzy time series systems:

1. Development of general concept of fuzzy time series forecasting systems by applying type-1 NSFLS, *i.e.*, by adding membership functions, non singleton fuzzification, inference engine and defuzzification to crisp FTS systems. Thus, a logical rule based type-1 NSFTS system is created.
2. By adding two other steps to the type-1 NSFTS system that was described in previous paragraph, a type-2 NSFTS system is built. First step is added after fuzzifier in fuzzy system that generates type-2 fuzzy sets from type-1 fuzzy sets by fuzzy normal forms and Yager [27] parametric classes of negation, T-norm and T-conorm. In second step type reduction is added before defuzzification in the output processing of the system.

This model uses the same number of fuzzy sets with the same rules structure as the type-1 NSFTS system. The only difference is that the antecedent and consequent are type-2 fuzzy sets and a type reduction is implemented before defuzzification. In addition, this model is capable to receive crisp, type-1, and type-2 input variables.

3. In order to minimize the effects of uncertainties and generating trained rules; the main parameters of the system are tuned by applying Genetic Algorithm. In this method, the parameters of fuzzy sets are tuned instead of the parameters of rule table. Then, the system automatically can tune type-2 fuzzy sets in shorter response time. This method leads to more accurate results than the model that the parameters of rules table are tuned.

This paper uses the version of two factors high order fuzzy time series that was presented by Lee *et al.* [9, 10] owing to its better capabilities among other methods. Finally for validation, the proposed methods are used for prediction of temperature and TAIEX index.

The rest of the paper is organized as follows: second section focuses on type-2 fuzzy sets and systems and their associated terminologies. Third section will review the initial concepts of type-1 Fuzzy Time Series. In fourth section, the proposed method for two factors, high order, non singletontype-1 and interval type-2 fuzzy logic system is presented. In fifth section fundamental concept of genetic algorithm is reviewed then designed approach to tuning the parameters of the system with genetic algorithm is presented. In sixth section, the proposed system for type-1 fuzzy time series is applied for forecasting temperature. In seventh section, the TAIEX stock index is predicted by both proposed type-1 and interval type-2 systems with genetic algorithm. Finally, conclusions and further researches are appeared in the last section.

Interval Type-2 Fuzzy Logic Sets and Systems

In this section, type-2 fuzzy sets and systems and their related terminologies used in this paper will briefly be explained.

Type-2 Fuzzy Logic Sets

Zadeh [29] was the first who introduced Type-2 Fuzzy Logic Sets (T2FLSs) (and even higher dimensions) as the extension of type- 1 fuzzy sets which are very useful in such circumstances where it is difficult to determine an exact MF for a FS [7]. Therefore, they can be used to handle these uncertainties. Later Mendel and John [16] proposed that there are four ways in an FLS in which uncertainty occurs where using type-2 fuzzy logic sets are more appropriate:

1. The words that were used in antecedents or consequents of rules have a variety of meanings for different people or in different situations.
2. Consequents obtained by a group of experts may often be different for the same rule, as the experts necessarily will not be in agreement.
3. Only noisy data are available for tuning (optimizing) the parameters of an IT2 FLS.
4. Noisy measurements in the FLS are available.

Type-2 fuzzy sets have grades of membership that are fuzzy. In other words, at each value of the primary variable (*e.g.* temperature, cloud density, Stock index) there is a value point of the primary membership where the each membership is a function (the secondary MF) as third dimension [15, 16]. Therefore, the MF of a type-2 fuzzy set has three dimensions, and is the new third dimension (secondary grades) which provides new degree of freedom for handling uncertainties [16]. These functions are embedded T2 FS which sits on a T1FSs and has a non-zero MF value that leads to construction of General T2 FSs [7].

In the literature, Interval Type-2 FLSs (IT2FLSs) is usually used in different practical models as a simple case of the general T2 FS in which all of its secondary MF values are equal to one so they only involve interval sets. So an IT2 FS is characterized and described totally by its FOU (Footprint of Uncertainty) [16] which can be observed in Fig. (**1**). A type-2 fuzzy set is described in following definitions:

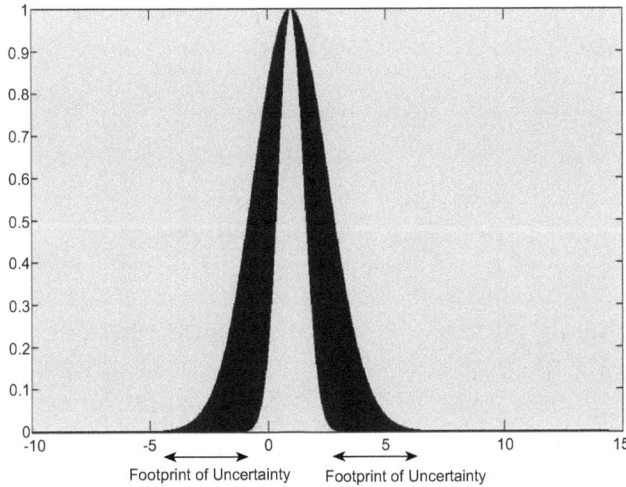

Fig. (1). Interval Type-2 fuzzy set with Gaussian MF and Uncertain Standard Deviation [17].

Definition 1: Mendel *et al.* [15], a T2FS, denoted $\tilde{A}$, is characterized by a type-2 MF $\mu_{\tilde{A}}(x,\mu)$, where $x \in X$ and $\mu \in J_x \subseteq [0,1]$ *i.e.,*

$$\tilde{A} = \{((x,\mu),\mu_{\tilde{A}}(x,\mu))|\forall x \in X, \forall \mu \in J_x \subseteq [0,1]\} \tag{1}$$

In which $0 \leq \mu_{\tilde{A}}(x,\mu) \leq 1$. $\tilde{A}$ can also be expressed as:

$$\tilde{A} = \int_{x\in X} \int_{\mu\in J_x} \mu_{\tilde{A}}(x,\mu)/(x,\mu)\, J_x \subseteq [0,1] \tag{2}$$

where, $\iint$ denotes union over all admissible x and μ. For a discrete universes of discourse, $\int$ is replaced by $\sum$. In this definition, the first restriction that $\forall \mu \in J_x \subseteq [0,1]$ is consistent with the T1 constraint that $0 \leq \mu_A(x) \leq 1$, *i.e.,* when uncertainties disappear a T2 MF must reduce to a T1 MF.

The second restriction that $0 \leq \mu_{\tilde{A}}(x,\mu) \leq 1$ is consistent with the fact that the amplitudes of a MF should lie between or be equal to 0 and 1 in which case the variable equals $\mu_A(x)$ and $0 \leq \mu_A(x) \leq 1$.

Definition 2: Mendel *et al.* [15], When all $\mu_{\tilde{A}}(x,\mu) = 1$, then $\tilde{A}$ is an IT2FS. Although the third dimension of the general T2FS is no longer needed because it conveys no new information about the IT2FS, the IT2FS can still be expressed as a special case of the general T2FS in (2), as:

$$\tilde{A} = \int_{x \in X} \int_{\mu \in J_x} 1/(x,\mu) J_x \subseteq [0,1] \tag{3}$$

Non Singleton Interval Type-2 Fuzzy Logic Systems

In order to design an FLS, inputs and outputs membership functions, overlapping between these functions, inference, implication, and aggregation methods and at last the defuzzification methods should be considered. Then we can design variety of FLSs [13]. Before explanation of IT2FLSs it should be considered that the difference between type-1 and type-2 FL systems is that at least one antecedent or one consequent set in a rule is made of type-2 fuzzy set [8, 15].

General structure of type-2 fuzzy logic system, which can be observed from Fig. (**2**) that contains five components: fuzzifier, inference engine, rule base and finally type reducer and defuzzifier should apply for computing crisp outputs [15]. This kind of FLS is widely used in many engineering applications of fuzzy logic.

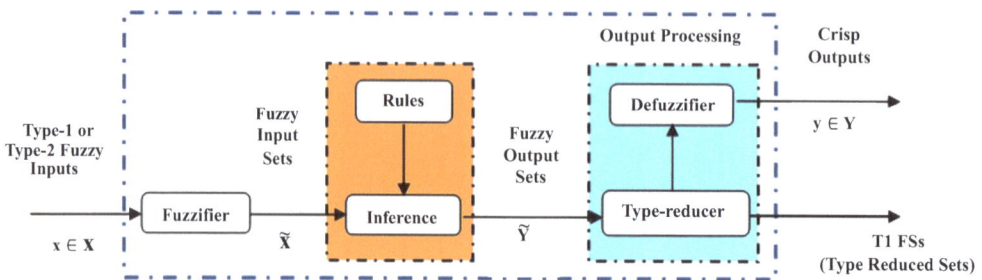

Fig. (2). Non singleton Type-2 fuzzy logic system [15].

There are two kinds of inputs in type-1 and type-2 fuzzy logic systems: crisp and type-1 inputs that led to singleton and non singleton fuzzifications respectively [13]. Non singleton fuzzification is especially useful in cases where the input data to the FLS contains uncertainty [13]. In this paper, we use the second approach for two reasons: firstly, it benefits from higher degree of accuracy, and, secondly, it does not only account for uncertainties of the antecedents or consequents in rules, but also accounts for input measurement uncertainties.

In this paper Mamdani inference engine with FITA (First Infer, Then Aggregate) approach [23] is applied. Moreover, parametric Yager and Filev's [26] defuzzification method based on the probabilistic nature of the selection process among the values of a fuzzy set, called Basic Defuzzification Distribution (BADD) method is used:

$$y(c) = \frac{\sum_{k=1}^{n} c^s(z)z}{\sum_{k=1}^{n} c^s(z)} \tag{4}$$

Determination of Footprints of Uncertainty (Umf and Lmf) in Interval Type-2 Fuzzy Logic Sets

IT2 fuzzy sets are characterized by their FOU, which are characterized by their boundaries, *i.e.*, Upper Membership Function (UMF) and Lower Membership Function (LMF) [17]. There are some methods for determination of UMF and LMF, which are described, in next paragraph.

Mendel [14] mentioned that there are two different approaches for selecting of type-2 FLS's parameters. One is the dependent approach in which the best possible type-1 FLS is designed, and then it is used to initialize the parameters of a type-2 FLS. The other method is the independent approach in which all the parameters of the type-2 FLS are tuned at the outset without the aid of an existing type-1 design. One advantage offered by the dependent approach is smarter initialization of the parameters of the type-2 FLS. Since the baseline type-1 fuzzy sets impose constraints on the type-2 sets, fewer parameters need to be tuned and the search space for each variable is smaller. Therefore, the computational cost is less than the independent approach. Yet both the FLSs have the same number of MFs. Thus in this system for increasing the robustness of the system, we use dependent approach.

Mendel & Wu [17, 18] have presented two methods for dependent approach for forward problems and inverse problems.

In this paper another dependent approach for transforming a type-1 fuzzy set to an interval type-2 fuzzy set is selected that was introduced by Turksen [22]. He proves that in analogy to the two valued set and logic theory where FDCF (Fuzzy Disjunctive Canonical Form) is equivalent to DNF (Disjunctive Normal Form) or FDCF(.) = DNF(.) and FCCF (Fuzzy Conjunctive Canonical Form) is equivalent to CNF (Conjunctive Normal Form) or FCCF(.) = CNF(.) inform only [23].

The equivalence, DNF(.) = CNF(.) breaks down, *i.e.*, we have FDCF(.) $\subseteq$ FCCF(.) for certain classes of t-norms and t-conorms that are strict and nilpotent Archimedean. FDCF and FCCF can be obtained for each combination of any two fuzzy sets where "AND", "OR" do not correspond to "∩" , "U" in a one to one mapping [23].

So by considering two linguistic concepts A and B, *i.e.*, the two predicates, words or fuzzy sets A and B then FOU of interval type-2 fuzzy set C is calculated as follows [23]:

$$
\begin{aligned}
C = A\ AND\ B &= \begin{cases} FDCF(AANDB) \\ FCCF(AANDB) \end{cases} \\
&= \begin{cases} A \cap B \\ (A \cup B) \cap (c(A) \cup B) \cap (A \cup c(B)) \end{cases} \\
&= \begin{cases} C_L & for\ Lower\ membership\ function \\ C_U & for\ Upper\ membership\ function \end{cases}
\end{aligned} \tag{5}
$$

In this paper for computing above formulas and making type-2 fuzzy sets, Yager [27] parametric classes of Complement, T-norm and T-conorm have been used which are as follows respectively:

Yager [27] complement:

$$ C_p(A) = (1 - A^p)^{1/p} \quad (p > 0) \tag{6} $$

Yager [27] T-norm:

$$ \left(1 - min\left(1, (1 - A)^q + (1 - B)^q\right)^{\frac{1}{q}}\right)\ for\ all\ A, B \in [0,1]\ and\ (q > 0) \tag{7} $$

Yager [27] T-conorm:

$$ min\left(1, (A^r + B^r)^{\frac{1}{r}}\right)\ for\ all\ A, B \in [0,1]\ and\ (r > 0) \tag{8} $$

Fundamental Concepts of Fuzzy Time Series

This section briefly reviews the concept of fuzzy series from Song and Chissom [20, 21] and Lee *et al.* [9, 10].

Definition 1: Song and Chissom [19], A fuzzy set *"A"* of the universe of discourse U = $\{u_1, u_2, ..., u_n\}$, is defined as follows: $A = f_A(u_1)/u_1 + ... + f_A(u_n)/u_n$ where $f_A(u_i)$, $i = 1,2, ...$ is the membership function of the fuzzy set *"A"*, f_A: $U \rightarrow$ [0,1], $f_A(u_i)$ denotes the grade of membership of u_i in the fuzzy set *"A"*.

Definition 2: Song and Chissom [19, 20], Let $Y(t)$, $t = 1,2, ...$ be a subset of R^1, the universe of discourse on fuzzy sets $f_i(t)$, $i = 1,2, ...$ is defined and $F(t)$ be a collection of $f_1(t), f_2(t), ...$ Then, $F(t)$ is a fuzzy time series which defined on $Y(t)$. If $F(t)$ only caused by $F(t-1)$, *i.e.*, $F(t-1) \rightarrow F(t)$, then it can be expressed as $F(t) = F(t-1) \circ R(t, t-1)$, where "$\circ$" represent the Max-Min composition operator. $R(t, t-1)$ is the fuzzy relationship between $F(t-1)$ and $F(t)$. Then $R(t, t-1)$ is named the first order forecasting model of $F(t)$. $F(t-1) \rightarrow F(t)$ is named a fuzzy logical relationship. If for any time t, $R(t, t-1)$ is independent of t, *i.e.*, for any time t, $R(t, t-1) = R(t, t-2)$ then $F(t)$ is named a time invariant fuzzy time series. Otherwise, it is named a time variant fuzzy time series.

Definition 3: Song and Chissom [19 - 21], Chen [1], Let $F(t)$ is a fuzzy time series. If $F(t)$ is caused by $F(t-n),...,F(t-1)$ then this Fuzzy Logical Relationship or Rule (FLR) is represented by $F(t-n),...,F(t-1) \rightarrow F(t)$ and it is named the k-order fuzzy time series forecasting model, where $n \geq 2$ and $F(t-n),...,F(t-1)$ and $F(t)$ are named the current state and the next state, respectively.

Definition 4: Lee *et al.* [9, 10], Let $F_1(t)$ and $F_2(t)$ be the first and second factor fuzzy time series respectively. If $F_1(t)$ is caused by $(F_1(t-1), F_2(t-1))$, $(F_1(t-n), F_2(t-n))$ then this Fuzzy Logical Rule is represented by $(F_1(t-n), F_2(t-n)),..., (F_1(t-1), F_2(t-1)) \rightarrow F_1(t)$ and it is named the two factors k order fuzzy time series forecasting model $k, t = 1,2,...$ and $(F_1(t-n), F_2(t-n)),...,(F_1(t-1), F_2(t-1))$ and $F(t)$ are named the current state and the next state, respectively. In these models, fuzzy time series has the same possible values.

Proposed Two Factors High Order Non Singleton Type-1 and Interval Type-2 Fuzzy Time Series Systems

This section, presents type-1 and interval type-2 fuzzy systems for forecasting fuzzy time series. These systems have been inspired from two factors high order fuzzy time series model which proposed by Lee *et al.* [9], using non singleton fuzzification, inference engine, type reduction and defuzzification. In addition, by replacing the consequents of rules with second factor, this factor can be predicted by the systems.

The steps of the proposed systems are as follows:

Step 1: The membership functions are assigned for the system that their parameters are calculated from historical data. In fact, there are varieties of membership functions and by choosing of different kinds of them, the system can calculate different results with the same parameters.

Step 2: Lee *et al.* [9], Define the universe of discourse U and V, where $U = [D_{min} - D_1, D_{max} + D_2]$ with D_{min} and D_{max} and $V = [E_{min} - E_1, E_{max} + E_2]$ with E_{min} and E_{max} that are the minimum and maximum values of the historical data of the first and second factor respectively. D_1, D_2, E_1 and D_2 are numbers to adjust the lower bound and the upper bound of the universe of discourse. Then divide the universe of discourse U and V into the same lengths of intervals u_1, u_2, ..., u_n and v_1, v_2, ..., v_l.

Step 3: Lee *et al.* [9], Fuzzify the historical data and define linguistic terms of the first and second factor into fuzzy sets, respectively. If the historical data of the first factor belongs to u_i where $1 \leq i \leq n$ and the historical data of the second factor belongs to v_i where $1 \leq j \leq l$, then fuzzify the historical data of the first and second factor into fuzzy set A_i and B_i respectively.

Step 4: The fuzzifier takes input values and determines the degree to which belong to each fuzzy set via membership functions by calculating their parameters. Define parameters of membership function of each linguistic term A_i and B_i represented by the interval u_i and v_i respectively. They are (m_{Ai}, σ_{Ai}) and (m_{Bi}, σ_{Bi}) that can be obtained from average of means and standard deviations of those historical data that are related to set A_i and B_i for the first and second factor respectively as follows:

$$A_1 = \frac{m_{A1}, \sigma_{A1}}{u_1}, A_2 = \frac{m_{A2}, \sigma_{A2}}{u_2}, A_3 = \frac{m_{A3}, \sigma_{A3}}{u_3}, ..., A_n = \frac{m_{An}, \sigma_{An}}{u_n} \qquad (9)$$

$$B_1 = \frac{m_{B1}, \sigma_{B1}}{v_1}, B_2 = \frac{m_{B2}, \sigma_{B2}}{v_2}, B_3 = \frac{m_{B3}, \sigma_{B3}}{v_3}, ..., B_m = \frac{m_{Bl}, \sigma_{Bl}}{v_l} \qquad (10)$$

Step 5: Transforming type-1 FLSs to interval type-2 FLSs with fuzzy normal forms by applying Yager [27] parametric classes of Complement, T-norm and T-conorm.

In this system, we should generate type-2 fuzzy sets from one linguistic concept so upper and lower membership function of each set of the first factor is calculated as follows:

$$\tilde{A}_i = \begin{cases} FDCF(A_i AND A_i) = \tilde{A}_{iL} \\ FCCF(A_i AND A_i) = \tilde{A}_{iU} \end{cases} \quad for\ i = 1,2,\dots,n \qquad (11)$$

where, $\tilde{A}_i$ is type-2 fuzzy set of the first factor. Upper and lower membership function of each individual set of the second factor is calculated as follows:

$$\tilde{B}_j = \begin{cases} FDCF\big(B_j AND B_j\big) = \tilde{B}_{jL} \\ FCCF\big(B_j AND B_j\big) = \tilde{B}_{jU} \end{cases} \quad for\ j = 1,2,\dots,l \qquad (12)$$

where, $\tilde{B}_j$ is type-2 fuzzy set of the second factor.

Note: As mentioned before, this method is capable of receiving crisp, type-1 and type-2 inputs so as long as type-2 inputs applied, step 5 can be skipped and in step 4, upper and lower bounds of type-2 fuzzy sets should be calculated.

Step 6: Lee *et al.* [9], as described in definition 4 of third section, based on the fuzzified first and second factors obtained in step4 two-factor k-order fuzzy logical rules are constructed.

Step 7: The two factors k-order fuzzy logical rules, which was generated in step 6 should be conformed to the following principles:

Principle 1. If some of the two factors k-order fuzzy logical rules have the same consequent, or $((A_{ik}, B_{jk}),\dots,(A_{i2}, B_{j2}),(A_{i1}, B_{j1})) \rightarrow A_{i1}, A_{i2},\dots,A_{itt}$ then the following order should apply in order to change them into MISO (Multiple Input single Output) rule so the consequent of the fuzzy logical rule is calculated as follows:

$\cap(A_{it},\dots,A_{i2},A_{i1}) = O$, thus the result is $(A_{ik}, B_{jk}),\dots,(A_{i2},B_{j2}),(A_{i1},B_{j1})) \rightarrow O$

In this case, we use classical T-norm (min). However, any kinds of T-norm in this principal can be used.

Principle 2. If there is a two factors k-order fuzzy logical rule that has no consequent, it should be omitted.

Principle 3. If there exists repeated or the same rules, keep one of them and omit the others as they have no effect on the inference of the system.

Step 8: In this step, inference is done.

Step 9: For computing crisp outputs in the output processing, a type reduction and defuzzification, method should be chosen to gain the centroid of aggregated rules.

Step 10: Evaluating performance of the system by comparing of forecasted data with actual data via Root Mean Square Error (RMSE) and Average Forecasting Error Rate (AFER). The smaller RMSE and AFER, results in higher accuracy.

$$AFER = \frac{\sum_{i=1}^{n} |(Forcasted\ value\ of\ Day\ i - Actual\ value\ of\ Day\ i)/Actual\ Value\ of\ Day\ i|}{n} * 100 \qquad (13)$$

$$RMSE = \sqrt{\frac{\sum_{i=1}^{n}(actual_i - forcasting_i)^2}{n}} \qquad (14)$$

where, n is the number of forecasted values.

Step 11: Optimize the main parameters of the system with genetic algorithm. Main parameters that need to be optimized are as follows:

A. Yager parameter of Complement, (p).
B. Yager parameter of T-norm, (q).
C. Yager parameter of T-conorm (S-norm), (r).
D. Parameters of first factor fuzzy sets (m_{Ai}, σ_{Ai}) where $1 \leq i \leq n$
E. Parameters of second factor fuzzy set (m_{Bj}, σ_{Bj}) where $1 \leq j \leq l$
F. Defuzzification parameter (s).

A genetic algorithm for tuning the main parameters of the system is proposed which is explained in the next section.

Tuning Method for Type-1 and Interval Type-2 FTSs with Genetic Algorithm

Following Darwin's principal of "Survival of the Fittest", evolutionary computational algorithms raised in order to optimize complicated problems. Genetic algorithms (GA) are a class of evolutionary computational algorithms, which are efficient optimization tools. GAs can optimize in a similar manner, by simulating the Darwinian evolutionary process and naturally occurring genetic operators on chromosomes [3, 5]. For solving problems in GA, first, a population of chromosomes is formed and each chromosome represents a possible solution to the problem.

In many researches, GAs have been successfully used to tune and design the MFs and the rules of FLSs. GA can optimize parameters of a fuzzy system representing the MFs and the rules of type-1 or type-2 FLSs in order to give better performance. For applying genetic algorithms, the feature parameters of a type-1 FLSs have to be encoded into a form of chromosome. There are various shapes of fuzzy sets such as Gaussian, Bell, Trapezoidal and Triangular. In each shape, several cases of parameters can be considered. In Gaussian encoding schemes for a type-1 FLS that we applied in this research, a type-1MF is represented as a mean (m) and a Standard Deviation (STD or σ) and rules represented as follows: for $l = 1, \ldots, M$.

$$R^l: IF x_1 \ isr \ \tilde{A}_1^l \ and \ \ldots x_p \ isr \ \tilde{A}_p^l \ THEN \ \tilde{Y}^l \Leftrightarrow \left(m_{x1}^l, \sigma_{x1}^l\right)\left(m_{x2}^l, \sigma_{x2}^l\right) \ldots \left(m_{xp}^l, \sigma_{xp}^l\right) \rightarrow \left(m_y^l, \sigma_y^l\right)$$

Many GA based FLS designing processes have been used to represent the rule table as genes. These methods make a rule based type-1 or type-2 FLS, as generating type-1 or type-2 rules including parameters of MFs.

In this research, another approach has been developed in which, parameters of type-1 fuzzy logic sets and parameters of operators, *i.e.*, Negation, T-norm and T-conorm and should be tuned. In this method instead of representing rules as gens, parameters of all fuzzy sets of the system will be represented as them. In addition, for reducing the response time, mean of each set is tuned in the scope of upper and lower bound of related interval that determined in step 2 of the proposed method.

Therefore, the domains of "m" (mean) for each set of the first and second factor are respectively:

$$u_1 \leq m_{A1} < u_2, \ldots, u_{n-1} \leq m_{Ai} \leq u_n \qquad 1 \leq i \leq n$$

$$v_1 \leq m_{B1} < v_2, \ldots, v_{l-1} \leq m_{Bj} \leq v_l \qquad 1 \leq j \leq l$$

The domains of "σ" (STD) for the first and second factor are $[0, \sigma_{Amax} + \alpha]$ and $[0, \sigma_{Bmax} + \beta]$, where "$\alpha$" and "$\beta$" are numbers to adjust the upper bounds of the STD of the first factor and second factor respectively.

The length of chromosomes in this approach is calculated as follows:

(Number of intervals * Number of Factors * Number of Parameters for each set) + Number of parameters of operators (complement, T-norm, T-conorm and Defuzzification).

In this approach type-1 rules that have been made from type-1fuzzy sets and corresponding parameters will be tuned. Consequently, Type-2 fuzzy sets that were made up of type-1 fuzzy sets by fuzzy normal forms will be tuned too. In addition, tuned fuzzy sets are usable in type-1 fuzzy time series forecasting system. Moreover, this approach gives results that are more accurate and less response time compared with those tuning methods that tune parameters of rule table.

GA's process fitness function for evaluation is one of the key factors to determine the performance of solutions and to control the speed of evolution in the genetic based FLSs. It is not easy to evaluate the performances of the FLSs so the results of executions are evaluated as a performance measure after simulating a designed FLS for given situations. It should be mentioned that to evaluate type-2 FLSs, fitness function considers the results of simulation executions by the RMSE between the training data and the crisp output (forecasted data) of a type-2 FLS.

Experimental Results by Temperature Prediction and TAIEX Forecasting

Temperature Prediction with Proposed Method

In this section, we apply the proposed two factors high order NS type-1 FTS system with two kinds of Gaussian membership functions (Table **1**). The proposed method is applied without tuning the main parameters of the system with genetic algorithm in order to show which of MFs give a higher degree of forecasting accuracy and how system computes better results than previous methods for prediction of the daily Temperature. In this example, temperature is used as the main factor and the daily Cloud density is used as the second factor. Table **2** shows the historical data and its parameters of the daily temperature and cloud density of June 1, 1998 to June 30, 1998 in Taipei, Taiwan (Central Weather Bureau, 1996).

Table 1. Different kinds of membership functions.

Types of MF	Formula
Gaussian MF	$u_i^l(x_i) = EXP\left(-\frac{1}{2}\left(\frac{x_i - m_i^l}{\sigma_i^l}\right)^2\right)$
Sharp Gaussian MF	$u_i^l(x_i) = \frac{1}{1 + \sigma_i^l(x_i - m_i^l)^2}$

Table 2. Historical data, mean and standard deviation of the daily temperature (UNIT: °C) and the cloud density of June 1996 in Taipei, Taiwan (Central Weather Bureau).

Date	JUNE Temp (m_i)	JUNE σ_i	JUNE Cloud Density (m_j)	JUNE σ_j	JULY Temp (m_i)	JULY σ_i	JULY Cloud Density (m_j)	JULY σ_j	AUGUST Temp (m_i)	AUGUST σ_i	AUGUST Cloud Density (m_j)	AUGUST σ_j	SEPTEMBER Temp (m_i)	SEPTEMBER σ_i	SEPTEMBER Cloud Density (m_j)	SEPTEMBER σ_j
1	26.09	1.66	36	3.50	30.00	2.61	15	3.68	27.13	1.36	100	2.75	27.58	1.55	29	2.78
2	27.55	2.11	23	1.13	28.42	2.69	31	2.63	28.96	2.65	78	3.26	26.88	1.42	53	3.04
3	29.03	3.37	23	1.88	29.38	2.63	26	2.92	29.04	1.81	68	3.11	26.42	0.91	66	3.43
4	30.52	3.91	10	2.49	29.46	2.02	34	4.10	29.29	2.34	44	3.60	27.50	1.76	50	2.8
5	30.01	3.12	13	3.77	29.96	2.89	24	2.75	28.88	2.11	56	3.77	26.63	1.07	53	2.83
6	29.45	1.95	30	3.36	29.63	2.84	28	3.45	28.67	1.65	89	3.81	28.29	1.90	63	2.54
7	29.71	2.10	45	1.98	30.21	2.60	50	2.48	29.00	1.87	71	3.24	29.25	2.26	36	1.49
8	29.41	1.66	35	1.39	29.33	2.19	34	3.43	28.17	2.13	28	3.42	29.04	1.43	76	3.01
9	28.79	1.94	26	2.87	28.08	1.32	15	3.21	27.13	2.11	70	3.96	30.29	2.17	55	2.01
10	29.41	2.29	21	3.13	28.92	2.27	8	1.97	28.33	2.66	44	3.37	30.04	1.90	31	2.74
11	29.26	1.91	43	3.04	28.42	2.58	36	3.82	29.04	2.28	48	2.73	30.00	2.33	31	2.52
12	28.53	2.01	40	2.74	29.75	3.55	13	3.01	28.17	2.19	76	2.81	30.67	1.52	25	2.17
13	28.69	1.80	30	2.56	27.88	3.24	26	3.68	29.92	2.18	50	2.85	30.29	2.23	14	1.87
14	27.53	2.67	29	4.02	29.13	3.10	44	3.46	27.67	2.54	84	4.21	30.29	2.28	45	1.85
15	29.49	3.09	30	3.58	27.67	2.49	25	3.67	26.83	2.37	69	2.87	29.63	2.12	38	2.2
16	28.81	3.06	46	3.43	28.04	2.92	24	2.87	27.63	2.50	78	2.90	28.33	1.55	24	2.84
17	29.02	2.99	55	2.60	28.83	3.21	26	3.51	27.96	2.54	39	2.63	28.71	2.07	19	1.9
18	30.31	2.60	19	2.60	29.96	3.05	25	3.64	29.08	2.45	20	3.71	28.21	2.90	39	1.14
19	30.19	2.72	15	3.59	30.79	2.18	21	4.14	29.25	2.42	24	2.80	28.38	2.93	14	1.16
20	30.93	2.86	56	2.64	31.71	2.76	35	2.29	29.83	2.41	25	2.30	28.38	2.18	3	4.09
21	30.83	2.79	60	3.44	31.42	2.36	29	1.97	29.67	2.78	19	3.08	26.46	0.91	38	3.97
22	28.68	1.36	96	2.30	31.29	2.39	48	3.58	29.46	1.55	46	3.52	25.75	1.83	70	3.3
23	27.81	2.24	63	3.06	31.29	2.46	53	3.67	28.04	2.15	41	3.62	25.08	1.22	71	2.87
24	27.35	2.35	28	3.02	31.33	1.80	44	2.68	28.33	2.41	34	3.41	27.13	1.94	70	3.09
25	27.67	2.76	14	3.60	28.92	1.04	100	2.42	28.58	2.38	29	3.13	25.83	2.25	40	3.67
26	27.05	2.80	25	4.29	28.04	0.68	100	1.94	28.79	2.29	31	2.98	26.46	1.83	30	2.38
27	28.43	2.80	29	3.51	28.67	1.18	91	0.95	29.08	2.50	41	3.47	25.63	1.84	34	0.49
28	27.78	2.39	55	3.50	28.08	2.50	84	0.83	27.75	2.70	14	3.75	24.33	0.62	59	0.4
29	29.03	2.99	29	3.24	29.29	1.67	38	1.47	26.25	2.28	28	2.94	23.29	0.79	83	0
30	30.20	2.42	19	3.28	28.00	1.22	46	2.21	26.04	1.97	33	3.90	23.71	1.37	38	0.92
31					26.96	1.24	95	1.34	27.71	3.16	26	2.96				

Step 1: In this step, mean(m) and standard deviation (σ) of historical data of temperature and cloud density measured based on observations of historical data during each day. See example in Appendix. Then results are shown in Table **2**. Then different kinds of Gaussian functions (Table **1**) have used in order to making membership functions.

Step 2: For the first factor, D_{min} = 26.09, D_{max} = 30.90, these numbers are resulted from the lowest and highest temperature of the month and D_1 = 0.09, D_2 = 0.1, so the universe of discourse is U = [26,31. For the second factor V_{min} = 10, V_{max} = 96 and E_1 = 0, E_2 = 0 hence, the universe of discourse is V = [10,96].

Then the universe of discourses U and V are divided into18 intervals with the same lengths, results is shown in Table **3**. Fig. (**3**) shows the first three type-1 fuzzy variables (inputs) of the system with Sharp Gaussian MF from 6/01/1998 to 6/3/1998 for both factors respectively which are extracted from Table **2**.

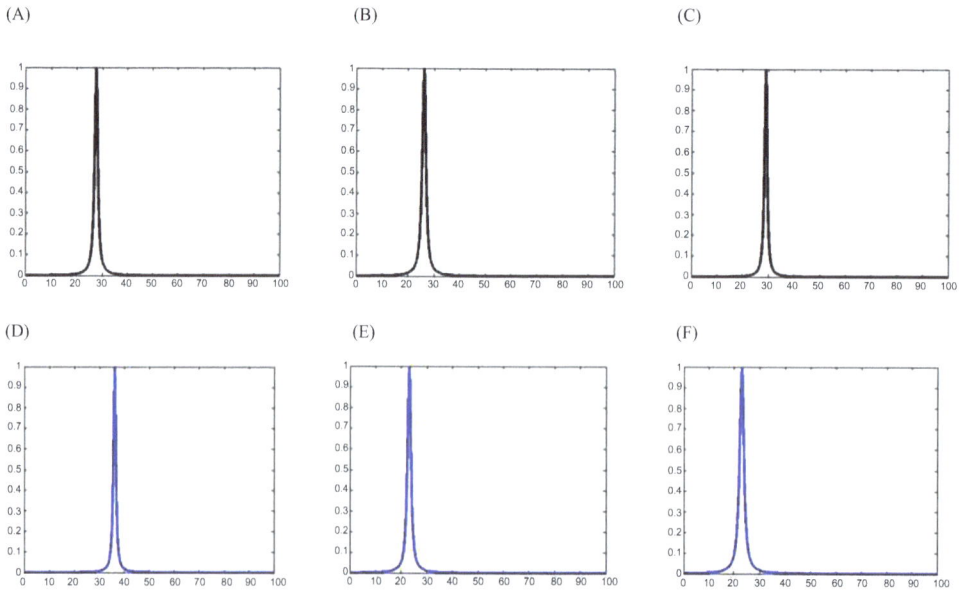

Fig. (3). Three type-1 fuzzy inputs of the system with Sharp Gaussian MF from 6/01/1998 to 6/3/1998 respectively for the first factor (A), (B), (C) and for the second factor (D), (E), (F).

Table 3. Upper and lower bounds of intervals of universe of discourse of the daily temperature and the cloud density for JUNE 1996 in Taipei Taiwan.

u_i	Left hand side u_{il}	Right hand side u_{ir}	v_j	Left hand side v_{jl}	Right hand side v_{jr}
u_1	26.00	26.28	v_1	10.00	14.78
u_2	26.28	26.56	v_2	14.78	19.56

(Table 3) contd.....

u_3	26.56	26.83	v_3	19.56	24.33
u_4	26.83	27.11	v_4	24.33	29.11
u_5	27.11	27.39	v_5	29.11	33.89
u_6	27.39	27.67	v_6	33.89	38.67
u_7	27.67	27.94	v_7	38.67	43.44
u_8	27.94	28.22	v_8	43.44	48.22
u_9	28.22	28.50	v_9	48.22	53.00
u_{10}	28.50	28.78	v_{10}	53.00	57.78
u_{11}	28.78	29.06	v_{11}	57.78	62.56
u_{12}	29.06	29.33	v_{12}	62.56	67.33
u_{13}	29.33	29.61	v_{13}	67.33	72.11
u_{14}	29.61	29.89	v_{14}	72.11	76.89
u_{15}	29.89	30.17	v_{15}	76.89	81.67
u_{16}	30.17	30.44	v_{16}	81.67	86.44
u_{17}	30.44	30.72	v_{17}	86.44	91.22
u_{18}	30.72	31.00	v_{18}	91.22	96.00

Step 3. Fuzzify the historical data of the first factor and second factor into fuzzy sets, respectively as shown in Table 4.

Table 4. Fuzzified historical data of the daily temperature and the cloud density for June 1996 in Taipei Taiwan.

Date	Temperature(m_i)	Fuzzified First Factor	Cloud Density Density(m_j)	Fuzzified Second Factor
6/01/1996	26.09	A_1	36	B_6
6/02/1996	27.55	A_6	23	B_3
6/03/1996	29.03	A_{11}	23	B_3
6/04/1996	30.52	A_{17}	10	B_1
6/05/1996	30.01	A_{15}	13	B_1
6/06/1996	29.45	A_{13}	30	B_5
6/07/1996	29.71	A_{14}	45	B_8
6/08/1996	29.41	A_{13}	35	B_6
6/09/1996	28.79	A_{11}	26	B_4
6/10/1996	29.41	A_{13}	21	B_3
6/11/1996	29.26	A_{12}	43	B_7
6/12/1996	28.53	A_{10}	40	B_7

(Table 4) contd.....

Date	Temperature(m_i)	Fuzzified First Factor	Cloud Density Density(m_j)	Fuzzified Second Factor
6/13/1996	28.69	A_{10}	30	B_5
6/14/1996	27.53	A_6	29	B_4
6/15/1996	29.49	A_{13}	30	B_5
6/16/1996	28.81	A_{11}	46	B_8
6/17/1996	29.02	A_{11}	55	B_{10}
6/18/1996	30.31	A_{16}	19	B_2
6/19/1996	30.19	A_{16}	15	B_2
6/20/1996	30.93	A_{18}	56	B_{10}
6/21/1996	30.83	A_{18}	60	B_{11}
6/22/1996	28.68	A_{10}	96	B_{18}
6/23/1996	27.81	A_7	63	B_{12}
6/24/1996	27.35	A_5	28	B_4
6/25/1996	27.67	A_7	14	B_1
6/26/1996	27.05	A_4	25	B_4
6/27/1996	28.43	A_9	29	B_4
6/28/1996	27.78	A_7	55	B_{10}
6/29/1996	29.03	A_{11}	29	B_4
6/30/1996	30.20	A_{16}	19	B_2

Step 4: The fuzzifier takes input values and determines the degree to which belong to each fuzzy set via membership functions by calculating their parameters, the results are shown in Tables **5** and **6**.

Table 5. Calculating Parameters of type-1 fuzzy sets of the daily temperature for June 1996 in Taipei Taiwan with Gaussian MF.

Means (m_i)	STDs (σ_i)	m_{Ai}	σ_{Ai}	Fuzzy Sets A_i
26.09	1.66	26.09	1.66	$A_1 = (26.9, 1.66)/u_1$
0	0	0.00	0.00	$A_2 = (0,0)/u_2$
0	0	0.00	0.00	$>A_3 = (0,0)/u_3$
27.05	2.80	27.05	2.80	$A_4 = (27.05, 2.8)/u_4$
27.35	2.35	27.35	2.35	$A_5 = (27.35, 2.35)/u_5$
27.55, 27.53	2.11, 2.67	27.54	2.39	$A_6 = (27.54, 2.39)/u_6$
27.81, 27.67, 27.78	2.24, 2.76, 2.39	27.75	2.46	$A_7 = (27.75, 2.46)/u_7$
0	0	0.00	0.00	$A_8 = (0,0)/u_8$

(Table 5) contd.....

Means (m_i)	STDs (σ_i)	m_{Ai}	σ_{Ai}	Fuzzy Sets A_i
28.43	2.80	28.43	2.80	$A_9 = (28.43,2.80)/u_9$
28.53, 28.69, 28.68	2.01, 1.8, 1.36	28.63	1.72	$A_{10} = (28.63,1.72)/u_{10}$
29.03, 28.79, 28.81, 29.02, 29.03	3.37, 1.94, 3.06, 2.99, 2.99	28.94	2.87	$A_{11} = (28.94,2.87)/u_{11}$
29.26	1.91	29.26	1.91	$A_{12} = (29.26,1.91)/u_{12}$
29.45, 29.41, 29.41, 29.49	1.95, 1.66, 2.29, 3.09	29.44	2.25	$A_{13} = (29.44,2.25)/u_{13}$
29.71	2.10	29.71	2.10	$A_{14} = (29.71,2.10)/u_{14}$
30.01	3.12	30.01	3.12	$A_{15} = (30.01,3.12)/u_{15}$
30.31, 30.19, 30.2	2.6, 2.72, 2.42	30.23	2.58	$A_{16} = (30.23,2.58)/u_{16}$
30.52	3.91	30.52	3.91	$A_{17} = (30.52,3.91)/u_{17}$
30.83, 30.93	2.86, 2.79	30.88	2.83	$A_{18} = (30.88,2.83)/u_{18}$

(Fig. **4**) shows the MFs of fuzzy sets A_i and B_j which extracted from last column of Table **5** and **6** for the first and second factor respectively.

Table 6. Calculating Parameters of type-1 fuzzy sets of the cloud density for JUNE 1996 in Taipei Taiwan with Gaussian MF.

Fuzzy Sets B_i	Means (m_j)	STDs(σ_j)	m_{Bj}	σ_{Bj}
$B_1 = (12.33,3.29)/v_1$	10, 13, 14	2.49, 3.77, 3.6	12.33	3.29
$B_2 = (17.67,3.16)/v_2$	15, 19, 19	2.6, 3.59, 3.28	17.67	3.16
$B_3 = (22.33,2.05)/v_3$	21, 23, 23	1.13, 1.88, 3.13	22.33	2.05
$B_4 = (27.67,3.49)/v_4$	26,29,28,25,29,29	2.87, 4.02,3.02,4.29,3.51,3.24,	27.67	3.49
$B_5 = (30,3.17)/v_5$	30,30,30	3.36,2.56,3.58	30.00	3.17
$B_6 = (35.5,2,45)/v_6$	36, 35	3.5, 1.39	35.50	2.45
$B_7 = (41.50,2.89)/v_7$	43, 40	3.04, 2.74	41.50	2.89
$B_8 = (45.5,2.71)/v_8$	45, 46	1.98, 3.43	45.50	2.71
$B_9 = (0,0)/v_9$	0	0	0.00	0.00
$B_{10} = (0,0)/v_{10}$	55,55,56	2.6,2.64,3.5	55.33	2.91
$B_{11} = (60,3.44)/v_{11}$	60	3.44	60.00	3.44
$B_{12} = (63,3.06)/v_{12}$	63	3.06	63.00	3.06
$B_{13} = (0,0)/v_{13}$	0	0	0.00	0.00
$B_{14} = (0,0)/v_{14}$	0	0	0.00	0.00
$B_{15} = (0,0)/v_{15}$	0	0	0.00	0.00
$B_{16} = (0,0)/v_{16}$	0	0	0.00	0.00
$B_{17} = (0,0)/v_{17}$	0	0	0.00	0.00
$B_{18} = (96,23)/v_{18}$	96	2.3	96.00	2.30

(A)   (B)

(C)   (D)

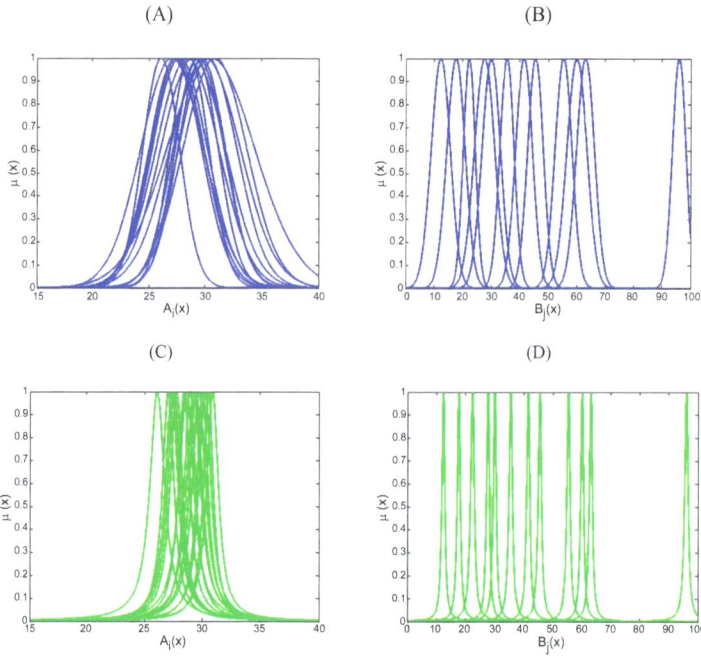

Fig. (4). (**A**) and (**B**) Fuzzy sets of first and second factor respectively with Gaussian MF; (**C**) and (**D**) Fuzzy sets of first and second factor respectively with Sharp Gaussian MF.

Step 5: This step is skipped as we used type-1 FTS system.

Step 6: Based on the fuzzified first factor and second factor historical data that obtained in Step 4, two factors k-order (k=3) fuzzy logical rules are generated as shown in Table **7**.

The MFs of the first rule is depicted in Fig. (**5**).

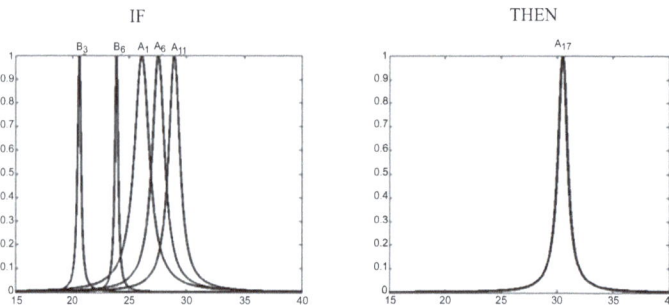

IF   THEN

Fig. (5). First rule with Sharp Gaussian MF.

Step 7: Based on Principle 2 the last rule is omitted.

Step 8: Inference the rules with FITA approach.

Table 7. Two factors three orders fuzzy logical rules of the data of June 1996.

Rule1:If (A_1,B_6) and (A_6,B_3) and (A_{11},B_3) Then A_{17}	Rule15: if(A_{13},B_5) and (A_{11},B_8) and $A_{11},B_{10})$ Then A_{16}
Rule2: if (A_6,B_3) and (A_{11},B_3) and (A_{17},B_1) Then A_{15}	Rule16: if (A_{11},B_8) and (A_{11},B_{10}) and (A_{16},B_2) Then A_{16}
Rule3: if (A_{11},B_3) and (A_{17},B_1) and (A_{15},B_1) Then iA_{13}	Rule17: if(A_{11},B_{10}) and (A_{16},B_2) and (A_{16},B_2) Then A_{18}
Rule4: if (A_{17},B_1) and (A_{15},B_1) and (A_{13},B_5) Then A_{14}	Rule18: if (A_{16},B_2) and (A_{16},B_2) and (A_{18},B_{10}) Then A_{18}
Rule5: if (A_{15},B_1) and (A_{13},B_5) and (A_{14},B_8) Then A_{13}	Rule19: if (A_{16},B_2) and (A_{18},B_{10}) and (A_{18},B_{11}) Then A_{10}
Rule6: if (A_{13},B_5) and $A_{14},B_8)$ and (A_{13},B_6) ThenA_{11}	Rule20: if (A_{18},B_{10}) and (A_{18},B_{11}) and (A_{10},B_{18}) Then A_7
Rule7: if (A_{14},B_8) and (A_{13},B_6) and (A_{11},B_4) Then A_{13}	Rule21: if (A_{18},B_{11}) and (A_{10},B_{18}) and (A_7,B_{12}) Then A_5
Rule8: if (A_{13},B_6) and (A_{11},B_4) and (A_{13},B_3) Then A_{12}	Rule22: if (A_{10},B_{18}) and (A_7,B_{12}) and (A_5,B_4) Then A_7
Rule9: if $(A_{11},B_4)$and (A_{13},B_3) and (A_{12},B_7) Then A_{10}	Rule23: if $(A_7,B_{12})$and (A_5,B_4) and (A_7,B_1) Then A_4
Rule10: if (A_{13},B_3) and (A_{12},B_7) and $A_{10},B_7)$ Then A_{10}	Rule24: if (A_5,B_4) and (A_7,B_1) and (A_4,B_4) Then A_9
Rule11: if (A_{12},B_7) and (A_{10},B_7) and (A_{10},B_5) Then A_6	Rule25: if (A_7,B_1) and (A_4,B_4) and (A_9,B_4) Then A_7
Rule12: if (A_{10},B_7) and (A_{10},B_5) and (A_6,B_4) Then A_{13}	Rule26: if (A_4,B_4) and (A_9,B_4) and (A_7,B_{10}) Then A_{11}
Rule13: if(A_{10},B_5) and $A_6,B_4)$ and (A_{13},B_5) Then A_{11}	Rule27: if $(A_9,B_4)$and (A_7,B_{10}) and (A_{11},B_4) Then A_{16}
Rule14: if (A_6,B_4) and (A_{13},B_5) and (A_{11},B_8) Then A_{11}	Rule28: if (A_7,B_{10}) and (A_{11},B_4) and (A_{16},B_2) Then #

Step 9: Defuzzify aggregated rules with parametric defuzzification in order to calculate forecasting data. Defuzzification parameter, $s = 7.01$.

Step 10: Evaluate the forecasting performance of NS type-1 FTS system which is defined by RMSE and AFER.

As can be observed in Table **8**, results and evaluation of the NS type-1 FTS system with Sharp Gaussian MF which has RMSE=0.06 and AFER=0.15%, is more accurate than the performance of the system with Gaussian MF which has RMSE=0.08 and AFER=0.21%.

In Fig. (**6**) the comparison between the forecasted and actual daily average temperature of June 1996 is shown which obtained by NS type-1 FTS system with Sharp Gaussian MF.

In this system, types of membership functions and operators, the length and number of interval in the universe of discourse and number of orders also affects the forecasting results.

Table 8. Forecasted Daily average temperature of June 1996 with different kinds of Gaussian membership functions with 18 fuzzy sets and three orders rules.

Date	Actual Temperature	Forecasted Temperature with Sharp Gaussian MF	Difference	Forecasted Temperature with Gaussian MF	Difference
6/04/1996	30.52	30.52	0.00	30.52	0.00
6/05/1996	30.01	30.01	0.00	30.01	0.00
6/06/1996	29.45	29.44	0.01	29.44	0.01
6/07/1996	29.71	29.71	0.00	29.71	0.00
6/08/1996	29.41	29.44	-0.03	29.44	-0.03
6/09/1996	28.79	28.94	-0.14	28.94	-0.14
6/10/1996	29.41	29.44	-0.03	29.44	-0.03
6/11/1996	29.26	29.26	0.00	29.45	-0.19
6/12/1996	28.53	28.63	-0.11	28.63	-0.11
6/13/1996	28.69	28.63	0.05	28.63	0.05
6/14/1996	27.53	27.54	-0.02	27.73	-0.20
6/15/1996	29.49	29.44	0.05	29.44	0.05
6/16/1996	28.81	28.94	-0.13	28.95	-0.14
6/17/1996	29.02	28.94	0.08	28.94	0.08
6/18/1996	30.31	30.23	0.07	30.23	0.08
6/19/1996	30.19	30.23	-0.04	30.23	-0.04
6/20/1996	30.93	30.88	0.05	30.87	0.06
6/21/1996	30.83	30.88	-0.05	30.88	-0.05
6/22/1996	28.68	28.63	0.05	28.63	0.05
6/23/1996	27.81	27.75	0.06	27.75	0.06
6/24/1996	27.35	27.35	0.00	27.35	0.00
6/25/1996	27.67	27.75	-0.09	27.75	-0.09
6/26/1996	27.05	27.05	0.00	27.05	0.00
6/27/1996	28.43	28.43	0.00	28.44	-0.01
6/28/1996	27.78	27.75	0.03	27.75	0.03
6/29/1996	29.03	28.94	0.09	28.94	0.09
6/30/1996	30.20	30.23	-0.03	30.23	-0.03
RMSE	**0.06**			**0.08**	
AFER	**0.15%**			**0.21%**	

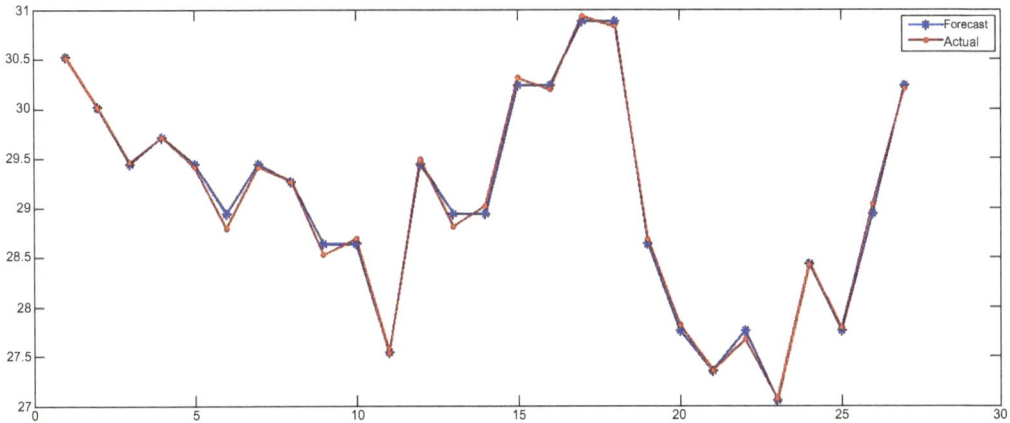

Fig. (6). Comparison between the forecasted and actual daily average temperature of June 1996 by using Sharp Gaussian membership function in the system.

The AFERs that obtained from different orders beside different number of sets with Sharp Gaussian and Gaussian MFs are shown in Tables **9** and **10** As shown in these tables, with 17 and 18 fuzzy sets (intervals) for all orders, the system results smaller number of AFERs, thus with more fuzzy sets (intervals) and more orders, higher degree of forecasting accuracy is confirmed.

Table 9. AFERs of June 1996 in Taipei Taiwan with different numbers of orders and fuzzy sets of the proposed method with sharp Gaussian MF.

Intervals	I=6	I=7	I=8	I=9	I=10	I=11	I=12	I=13	I=14	I=15	I=16	I=17	I=18
3Order	0.95%	0.84%	0.77%	0.44%	0.54%	0.57%	0.25%	0.39%	0.20%	0.19%	0.24%	0.29%	0.15%
4Order	0.77%	0.55%	0.63%	0.33%	0.38%	0.33%	0.24%	0.29%	0.19%	0.20%	0.20%	0.14%	0.16%
5Order	0.73%	0.55%	0.38%	0.32%	0.37%	0.32%	0.25%	0.28%	0.18%	0.19%	0.21%	0.14%	0.17%

Table 10. AFERs of June 1996 in Taipei Taiwan with different numbers of orders and fuzzy sets of the proposed method with Gaussian MF.

Intervals	I=6	I=7	I=8	I=9	I=10	I=11	I=12	I=13	I=14	I=15	I=16	I=17	I=18
3Order	0.92%	0.91%	0.85%	0.39%	0.70%	0.59%	0.28%	0.36%	0.37%	0.17%	0.22%	0.25%	0.21%
4Order	0.77%	0.58%	0.62%	0.33%	0.43%	0.33%	0.23%	0.29%	0.19%	0.20%	0.21%	0.15%	0.16%
5Order	0.73%	0.56%	0.38%	0.32%	0.39%	0.32%	0.25%	0.28%	0.18%	0.19%	0.21%	0.14%	0.17%

In Tables **11** to **14**, the comparison between AFERs of the proposed method with 18 fuzzy sets (intervals) and the methods that were presented by, Chen & Hwang [2], Lee *et al.* [10, 11], Wang & Chen [24] and Hsu *et al.* [4] are shown. The results show the higher degree of accuracy of the model for all orders above the previous methods.

Table 11. Comparison of the forecasting temperature for June 1996 for different methods.

Month June Order	Order 2	Order 3	Order 4	Order 5	Order 6	Order 7	Order 8
Chen and Hwang (2000)(window basis)	2.88%	3.16%	3.24%	3.33%	3.39%	3.53%	3.67%
Lee *et al.* (2006)	0.80%	0.76%	0.79%	0.76%	0.79%	0.79%	0.81%
Lee *et al.* (2007)	0.74%	0.64%	0.72%	0.65%	0.66%	0.64%	0.65%
Lee *et al.* (2008)	0.44%	0.42%	0.42%	0.42%	0.44%	0.40%	0.40%
Wang, and Chen (2009)	0.28%	0.29%	0.30%	0.29%	0.29%	0.28%	0.29%
Hsu et al (2010)	0.36%	0.36%	0.33%	0.32%	0.33%	0.28%	0.30%
Proposed Method with Sharp Gaussian MF	0.25%	0.15%	0.16%	0.17%	0.17%	0.18%	0.18%
Proposed Method with Gaussian MF	0.61%	0.21%	0.16%	0.17%	0.17%	0.18%	0.18%

Table 12. Comparison of the forecasting temperature for July 1996 for different methods.

Month July Order	Order 2	Order 3	Order 4	Order 5	Order 6	Order 7	Order 8
Chen and Hwang (2000)(window basis)	3.04%	3.76%	4.08%	4.17%	4.35%	4.38%	4.56%
Lee *et al.* (2006)	0.96%	0.96%	0.98%	0.97%	1%	0.98%	0.99%
Lee *et al.* (2007)	0.78%	0.73%	0.83%	0.70%	0.71%	0.68%	0.69%
Lee *et al.* (2008)	0.45%	0.42%	0.41%	0.41%	0.40%	0.41%	0.40%
Wang, and Chen (2009)	0.34%	0.35%	0.34%	0.34%	0.35%	0.33%	0.32%
Hsu et al (2010)	0.36%	0.35%	0.36%	0.34%	0.34%	0.36%	0.34%
Proposed Method with Sharp Gaussian MF	0.42%	0.28%	0.19%	0.19%	0.18%	0.17%	0.17%
Proposed Method with Gaussian MF	0.99%	0.44%	0.28%	0.18%	0.18%	0.17%	0.17%

Table 13. Comparison of the forecasting temperature for August 1996 for different methods.

Month August							
Order	Order 2	Order 3	Order 4	Order 5	Order 6	Order 7	Order 8
Chen and Hwang (2000)(window basis)	2.75%	2.77%	3.30%	3.40%	3.18%	3.15%	3.19%
Lee *et al.* (2006)	1.07%	1.06%	1.08%	1.08%	1.09%	1.07%	1.07%
Lee *et al.* (2007)	0.92%	0.88%	1.07%	0.75%	0.76%	0.75%	0.73%
Lee *et al.* (2008)	0.43%	0.47%	0.40%	0.41%	0.38%	0.40%	0.45%
Wang, and Chen (2009)	0.23%	0.22%	0.22%	0.22%	0.23%	0.23%	0.22%
Hsu et al (2010)	0.32%	0.35%	0.34%	0.33%	0.34%	0.35%	0.36%
Proposed Method with Sharp Gaussian MF	0.09%	0.10%	0.10%	0.09%	0.08%	0.08%	0.08%
Proposed Method with Gaussian MF	0.63%	0.32%	0.12%	0.09%	0.08%	0.08%	0.08%

Table 14. Comparison of the forecasting temperature for September 1996 for different methods.

Month September							
Order	Order 2	Order 3	Order 4	Order 5	Order 6	Order 7	Order 8
Chen and Hwang (2000)(window basis)	3.29%	3.10%	3.19%	3.22%	3.39%	3.38%	3.29%
Lee *et al.* (2006)	1.01%	0.9%	0.94%	0.96%	0.95%	0.95%	0.92%
Lee *et al.* (2007)	0.91%	0.86%	1.03%	0.87%	0.97%	0.84%	0.82%
Lee *et al.* (2008)	0.58%	0.59%	0.57%	0.56%	0.57%	0.58%	0.47%
Wang, and Chen (2009)	0.51%	0.49%	0.51%	0.51%	0.53%	0.5%	0.51%
Hsu et al (2010)	0.55%	0.57%	0.54%	0.51%	0.52%	0.53%	0.45%
Proposed Method with Sharp Gaussian MF	0.16%	0.17%	0.17%	0.15%	0.16%	0.14%	0.14%
Proposed Method with Gaussian MF	1.37%	0.70%	0.72%	0.73%	0.76%	0.83%	0.87%

TAIEX Forecasting By Applying the Proposed Method with Genetic Algorithm

In this section, both proposed NS type-1 FTS system and NSIT2 FTS system with Genetic Algorithm applied to predict the daily TAIEX in order to show how IT2FLS improves the results with a higher degree of accuracy than NS type-1 FTS system. In this example, the TAIEX is used as the first factor and the daily

TAIFEX (Taiwan Futures Exchange) is used as the second factor.

Moreover, the proposed NS type-1 FTS system has been applied to show how system can predict second factor (TAIFEX) by modifying the consequents of the rules with this factor.

Step 1: In this example, sharp Gaussian membership function has chosen, owing to the higher degree of accuracy. In this step, indexes considered as their mean for each historical data and standard deviation (σ) of each historical data measured based on its observations during each day for making type-1 fuzzy input variables.

Table **15** shows the historical data of the TAIEX and TAIFEX of November 1, 2000 to November 30, 2000 with their standard deviations (σ).

Table 15. Historical data, mean and standard deviation of the daily TAIEX and TAIFEX for November 2000.

Date	TAIEX(m_i)	σ_i	TAIFEX(m_j)	σ_j
11/01/2000	5425.02	30.76	5448	130
11/02/2000	5626.08	55.42	5797	168.50
11/03/2000	5796.08	37.25	5928	70.08
11/04/2000	5677.30	46.91	5722	119.92
11/06/2000	5657.48	20.13	5728	49.80
11/07/2000	5877.77	35.76	5987	103.19
11/08/2000	6067.94	81.28	6150	171.28
11/09/2000	6089.55	40.55	6215	97.26
11/10/2000	6088.74	23.09	6210	47.43
11/13/2000	5793.52	30.77	5800	110.05
11/14/2000	5772.51	29.9	5870	34.76
11/15/2000	5737.02	58.21	5731	157.25
11/16/2000	5454.13	35.48	5374	127.97
11/17/2000	5351.36	45.04	5330	62.55
11/18/2000	5167.35	49.27	5130	106.07
11/20/2000	4845.21	50.50	4771	148.06
11/21/2000	5103.00	83.21	5104	191.87
11/22/2000	5130.61	44.74	5039	111.01
11/23/2000	5146.92	25.07	5224	127.19
11/24/2000	5419.99	40.65	5510	131.81
11/27/2000	5433.78	29.35	5480	68.88

(Table 15) contd.....

Date	TAIEX(m_i)	σ_i	TAIFEX(m_j)	σ_j
11/28/2000	5362.26	14.64	5305	47.76
11/29/2000	5319.46	27.09	5338	80.84
11/30/2000	5256.93	38.30	5191	91.770

Step 2: Define the universe of discourse U and V and partition it into 16 intervals with equal lengths. Results are shown in Table **16**.

Table 16. Upper and lower bounds of intervals of universe of discourse of the daily TAIEX and TAIFEX for November 2000.

u_i	Left hand side u_{il}	Right hand side u_{ir}	v_j	Left hand side v_{jl}	Right hand side v_{jr}
u_1	4845.00	4922.81	v_1	4711.00	4810.38
u_2	4922.81	5000.63	v_2	4810.38	4909.75
u_3	5000.63	5078.44	v_3	4909.75	5009.13
u_4	5078.44	5156.25	v_4	5009.13	5108.50
u_5	5156.25	5234.06	v_5	5108.50	5207.88
u_6	5234.06	5311.88	v_6	5207.88	5307.25
u_7	5311.88	5389.69	v_7	5307.25	5406.63
u_8	5389.69	5467.50	v_8	5406.63	5506.00
u_9	5467.50	5545.31	v_9	5506.00	5605.38
u_{10}	5545.31	5623.13	v_{10}	5605.38	5704.75
u_{11}	5623.13	5700.94	v_{11}	5704.75	5804.13
u_{12}	5700.94	5778.75	v_{12}	5804.13	5903.50
u_{13}	5778.75	5856.56	v_{13}	5903.50	6002.88
u_{14}	5856.56	5934.38	v_{14}	6002.88	6102.25
u_{15}	5934.38	6012.19	v_{15}	6102.25	6201.63
u_{16}	6012.19	6090.00	v_{16}	6201.63	6301.00

Step 3: Fuzzify the historical data of the first factor and second factor into fuzzy sets, results are shown in Table **17**.

Table 17. Fuzzified historical data of the daily TAIEX and TAIFEX for November 2000.

Date	TAIEX(m_i)	Fuzzified First Factor	TAIFEX(m_j)	Fuzzified Second Factor
11/01/2000	5425.02	A_8	5448	B_8
11/02/2000	5626.08	A_{11}	5797	B_{11}
11/03/2000	5796.08	A_{13}	5928	B_{13}

(Table 17) contd.....

Date	TAIEX(m_i)	Fuzzified First Factor	TAIFEX(m_j)	Fuzzified Second Factor
11/04/2000	5677.30	A_{11}	5722	B_{11}
11/06/2000	5657.48	A_{11}	5728	B_{11}
11/07/2000	5877.77	A_{14}	5987	B_{13}
11/08/2000	6067.94	A_{16}	6150	B_{15}
11/09/2000	6089.55	A_{16}	6215	B_{16}
11/10/2000	6088.74	A_{16}	6210	B_{16}
11/13/2000	5793.52	A_{13}	5800	B_{11}
11/14/2000	5772.51	A_{12}	5870	B_{12}
11/15/2000	5737.02	A_{12}	5731	B_{11}
11/16/2000	5454.13	A_8	5374	B_7
11/17/2000	5351.36	A_7	5330	B_7
11/18/2000	5167.35	A_5	5130	B_5
11/20/2000	4845.21	A_1	4771	B_1
11/21/2000	5103.00	A_4	5104	B_4
11/22/2000	5130.61	A_4	5039	B_4
11/23/2000	5146.92	A_4	5224	B_6
11/24/2000	5419.99	A_8	5510	B_9
11/27/2000	5433.78	A_8	5480	B_8
11/28/2000	5362.26	A_7	5305	B_6
11/29/2000	5319.46	A_7	5338	B_7
11/30/2000	5256.93	A_6	5191	B_5

Step 4: Determine the degree that each value belongs to each fuzzy set via membership functions. Results are shown in Tables **18** and **19**.

Table 18. Calculating Parameters of type-1 fuzzy sets of the daily TAIEX for November 2000.

Means	Standard Deviations	m_{Ai}	σ_{Ai}	Fuzzy Sets of First Factor A_i
4845.21	50.50	4845.21	50.50	$A_1 = (4845.21,50.50)/u_1$
0	0	0.00	0.00	$A_2 = (0,0)/u_2$
0	0	0.00	0.00	$A_3 = (0,0)/u_3$
5103, 5146.92, 5130.61	83.21, 25.07, 44.74	5126.84	51.01	$A_4 = (5126.84,51.01)/u_4$
5167.35	49.27	5167.35	49.27	$A_5 = (5167.35,49.27)/u_5$
5256.93	38.30	5256.93	38.30	$A_6 = (5256.93,38.30)/u_6$
5351.36, 5362.26, 5319.46	27.09, 45.04, 14.64	5344.36	28.92	$A_7 = (5344.36,28.92)/u_7$

(Table 18) contd.....

Means	Standard Deviations	m_{Ai}	σ_{Ai}	Fuzzy Sets of First Factor A_i
5425.02, 5454.13, 5419.99, 5433.78	30.76, 35.48, 40.65, 29.35	5433.23	34.06	$A_8 = (5433.230, 34.060)/u_8$
0	0	0.00	0.00	$A_9 = (0,0)/u_9$
0	0	0.00	0.00	$A_{10} = (0,0)/u_{10}$
5677.30, 5657.48, 5626.08	46.91, 20.13, 55.42	5653.62	40.82	$A_{11} = (5653.62, 40.82)/u_{11}$
5772.51, 5737.02	29.99, 58.21	5754.77	44.10	$A_{12} = (5754.77, 44.10)/u_{12}$
5796.08, 5793.52	37.25, 30.77	5794.80	34.01	$A_{13} = (5794.8, 34,01)/u_{13}$
5877.77	35.76	5877.77	35.76	$A_{14} = (5877.77, 35.76)/u_{14}$
0	0	0.00	0.00	$A_{15} = (0,0)/u_{15}$
6089.55, 6088.74, 6067.94	40.55, 23.09, 81.28	6082.08	48.31	$A_{16} = (6082.08, 48.31)/u_{16}$

Table 19. Calculating Parameters of type-1 fuzzy sets of the daily TAIFEX for November 2000.

Means (m_j)	Standard Deviations(σ_j)	m_{Bj}	σ_{Bj}	Fuzzy Sets of Second factor B_i
4771	148.06	4771.00	148.06	$B_1 = (4771, 148.060)/v_1$
0	0	0.00	0.00	$B_2 = (0,0)/v_2$
0	0	0.00	0.00	$B_3 = (0,0)/v_3$
5104, 5039	191.87, 111.01	5071.50	151.44	$B_4 = (5071.5, 151.44)/v_4$
5130, 5191	106.07, 91.77	5160.50	98.92	$B_5 = (5160.5, 98.92)/v_5$
5305, 5224	47.76, 127.19	5264.50	87.48	$B_6 = (5264.5, 87.48)/v_6$
5330, 5338, 5374	80.84, 62.55, 127.97	5347.33	90.45	$B_7 = (5347.33, 90.45)/v_7$
5448, 5480	68.88, 130.64	5464.00	99.76	$B_8 = (5464, 99.76)/v_8$
5510	131.81	5510.00	131.81	$B_9 = (5510, 131, 81)/v_9$
0	0	0.00	0.00	$B_{10} = (0,0)/v_{10}$
5722,5728,5731,5797,5800	119.92,49.80,157.25,168.50,110.05	5755.60	121.10	$B_{11} = (5755.6, 121.1)/v_{11}$
5870	34.76	5870.00	34.76	$B_{12} = (5870, 34.76)/v_{12}$
5928,5987	70.08, 103.19	5957.50	86.64	$B_{13} = (5957.50, 86.64)/v_{13}$
0	0	0.00	0.00	$B_{14} = (0,0)/v_{14}$
6150	171.280	6150.00	171.28	$B_{15} = (6150, 171.28)/v_{15}$
6215, 6210	97.26, 47.43	6212.50	72.35	$B_{16} = (6212.50, 72.345)/v_{16}$

Step 5: Transformation type-1 FLSs to interval type-2 FLSs with fuzzy normal forms.

Step 6: Based on the fuzzified first factor and the fuzzified second factor obtained in Step 4, generate two factor k-orders (k=3) fuzzy logical rules. Results are shown in Table **20**.

Table 20. Two factors three orders fuzzy logical rules of data of November 2000.

Rule1: If (A_{11},B_{11}) and (A_8,B_8) and (A_{13},B_{13}) Then A_{11}
Rule2: If (A_{11},B_{11}) and (A_{13},B_{13}) and (A_{11},B_{11}) Then A_{11}
Rule3: If (A_{13},B_{13}) and (A_{11},B_{11}) and (A_{11},B_{11}) Then A_{14}
Rule4: If (A_{11},B_{11}) and (A_{11},B_{11}) and (A_{14},B_{13}) Then A_{16}
Rule5: If (A_{11},B_{11}) and (A_{14},B_{13}) and (A_{16},B_{15}) Then A_{16}
Rule6: If (A_{14},B_{13}) and (A_{16},B_{15}) and (A_{16},B_{16}) Then A_{16}
Rule7: If (A_{16},B_{15}) and (A_{16},B_{16}) and (A_{16},B_{16}) Then A_{13}
Rule8: If (A_{16},B_{16}) and (A_{16},B_{16}) and (A_{13},B_{11}) Then A_{12}
Rule9: If (A_{16},B_{16}) and (A_{12},B_{11}) and (A_{12},B_{12}) Then A_{12}
Rule10: If (A_{13},B_{11}) and (A_{12},B_{12}) and (A_{12},B_{11}) Then A_8
Rule11: If (A_{12},B_{12}) and (A_{12},B_{11}) and (A_8,B_7) Then A_7
Rule12: If (A_{12},B_{11}) and (A_8,B_7) and (A_7,B_7) Then A_5
Rule13: If (A_8,B_7) and (A_7,B_7) and (A_5,B_5) Then A_1
Rule14: If (A_7,B_7) and (A_5,B_5) and (A_1,B_1) Then A_4
Rule15: If (A_5,B_5) and (A_1,B_1) and (A_4,B_4) Then A_4
Rule16: If (A_1,B_1) and (A_4,B_4) and (A_4,B_4) Then A_4
Rule17: If (A_4,B_4) and (A_4,B_4) and (A_4,B_6) Then A_8
Rule18: If (A_4,B_4) and (A_4,B_6) and (A_8,B_9) Then A_8
Rule19: If (A_4,B_6) and (A_8,B_9) and (A_8,B_8) Then A_7
Rule20: If (A_8,B_9) and (A_8,B_8) and (A_7,B_6) Then A_7
Rule21: If (A_8,B_8) and (A_7,B_6) and (A_7,B_7) Then A_6
Rule22: If (A_7,B_6) and (A_7,B_7) and (A_6,B_5) Then #

Step 7: The last rule omitted based on principal 2.

Step 8: Inference the rules and inputs with FITA approach.

Step 9: In this step the defuzzification parameter, $s = 1.95$.

Step 10: Evaluate the forecasting performance is defined by RMSE and AFER.

Results of the evaluation of the proposed non singleton type-1 FTS system has RMSE=14.32 and AFER=0.22% for the forecasted daily TAIEX of November

2000, which is much less than Huarng [6] method with RMSE=160.94 and AFER=2.23%. Results are shown in Table **21**. Fig. (**7**) shows the comparison between the forecasted and actual daily TAIEX of November 2000.

Table 21. Forecasted daily average of TAIEX and TAIFEX for November 2000 with Sharp Gaussian MF.

Date	Actual TAIEX	Forecasted TAIEX By T1FTSs	Difference	Actual TAIFEX	Forecasted TAIFEX By T1FTSs	Difference
11/04/2000	5677.30	5653.76	23.54	5722	5755.35	33.35
11/06/2000	5657.48	5639.85	17.63	5728	5722.42	-5.58
11/07/2000	5877.77	5872.61	5.16	5987	5949.81	-37.19
11/08/2000	6067.94	6078.41	-10.47	6150	6143.97	-6.03
11/09/2000	6089.55	6081.11	8.44	6215	6211.62	-3.38
11/10/2000	6088.74	6080.61	8.13	6210	6210.79	0.79
11/13/2000	5793.52	5794.83	-1.31	5800	5755.71	-44.29
11/14/2000	5772.51	5754.62	17.89	5870	5869.83	-0.17
11/15/2000	5737.02	5755.44	-18.42	5731	5758.11	27.11
11/16/2000	5454.13	5476.13	-22.00	5374	5423.45	49.45
11/17/2000	5351.36	5344.45	6.91	5330	5347.47	17.47
11/18/2000	5167.35	5167.83	-0.48	5130	5160.99	30.99
11/20/2000	4845.21	4851.47	-6.26	4771	4779.07	8.07
11/21/2000	5103.00	5126.87	-23.87	5104	5071.52	-32.48
11/22/2000	5130.61	5127.40	3.21	5039	5072.34	33.34
11/23/2000	5146.92	5127.41	19.51	5224	5264.63	40.63
11/24/2000	5419.99	5431.87	-11.88	5510	5506.80	-3.20
11/27/2000	5433.78	5432.61	1.17	5480	5462.90	-17.10
11/28/2000	5362.26	5343.63	18.63	5305	5264.21	-40.79
11/29/2000	5319.46	5341.47	-22.01	5338	5343.22	5.22
11/30/2000	5256.93	5255.69	1.24	5191	5161.66	-29.34
RMSE	14.32			27.36		
AFER	0.22%			0.41%		

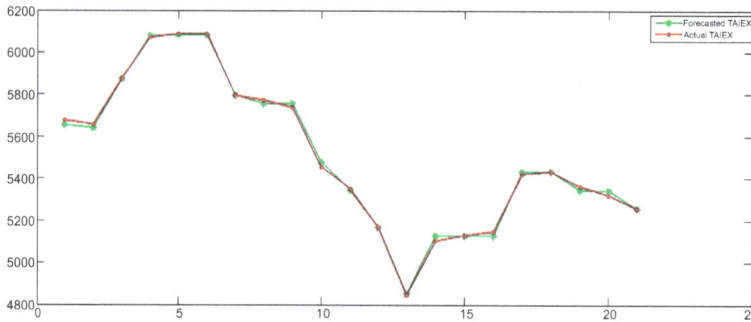

Fig. (7). Forecasted and actual daily TAIEX of November year 2000 byNS Type-1 FTS System withSharp Gaussian membership function.

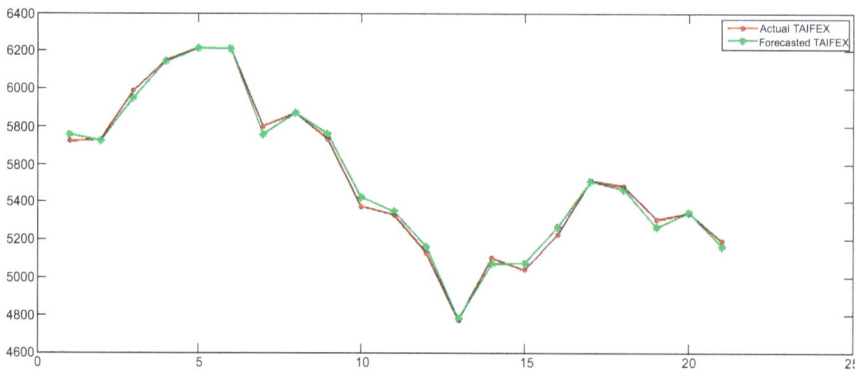

Fig. (8). Forecasted and actual daily TAIFEX of November year 2000 by NS Type-1 FTS System with Sharp Gaussian membership function.

This system is capable of predicting TAIFEX by replacing consequents with this factor. Forecasted values of the TAIFEX are shown in column 6 of Table **21** with RMSE=27.36 and AFER=0.41%. Fig. (**8**) shows the comparison between the forecasted and actual daily TAIFEX of November 2000.

Step 11: Tune the main parameters of the system by proposed Genetic Algorithm.

GA Procedure

To begin the GA, define 50 random initial populations. The parameters of each MF for sets are selected randomly with uniform probability density. Two parameters of each fuzzy set m and σ are chosen. Each chromosome is initialized for containing 32 random sets, 16 for first factor and 16 for second factor. The length of each chromosome is $16*2*2+4=68$ that means all parameters of the system equal 68. The domains of σ (STD) for the first and second factors are [0,90] and [0,200] with $\alpha = 8.72$ and $\beta = 8.13$ respectively and the domains of m

(mean) for each set of the first and second factor are according to Table **16**.

Selection and Pairing

Uniform selection used for choosing the parents.

Crossover

Crossover points are randomly selected, and then the variables in between of them are exchanged. Here single point crossover with Crossover fraction 0.7 used that specifies the fraction of the next generation.

Mutation and Reinsertion

The mutation rate is $\mu = 20\%$. To maintain the size of the original population, the elitist replacement strategy is used to insert the new individuals into the old population.

Termination Condition

Two constraints for termination of proposed algorithm have used, the GA will stop after 200 numbers of generations or the algorithm stops if cumulative change in the objective function for 50 generations is less than 1^{-10}.

Type Reduction and Defuzzification

Main parameters of the system are tuned by applying the proposed genetic algorithm. Results are shown in Table **22**. Fig. (**9**) shows the best mean RMSE of 200 generations and the best variables of the last generation.

Table 22. Tuned parameters of the system which obtained by the proposed genetic algorithm.

TAIEX Mean	TAIEX STD	Fuzzy Sets of First Factor A_i	TAIFEX Mean	TAIFEX STD	Fuzzy Sets of Second factor B_i
4836.49	0.12	$A_1 = (4836.49, 0.12)/u_1$	4788.49	0.46	$B_1 = (4788.49.0.46)/v_1$
4934.34	0.01	$A_2 = (4934.34, 0.01)/u_2$	4841.65	0.43	$B_2 = (4841.65, 0.43)/v_2$
5042.62	0.28	$A_3 = (5042.62, 0.28)/u_3$	4978.74	0.78	$B_3 = (4978.74, 0.78)/v_3$
5121.69	0.40	$A_4 = (5121.69, 0.40)/u_4$	5037.39	0.43	$B_4 = (5037.39.0.43)/v_4$
5167.20	0.31	$A_5 = (5167.20, 0.31)/u_5$	5108.20	0.39	$B_5 = (5108.20.0.39)/v_5$
5230.76	0.94	$A_6 = (5230.76, 0.94)/u_6$	5242.24	0.88	$B_6 = (5242.24.0.88)/v_6$
5355.31	0.24	$A_7 = (5355.31, 0.24)/u_7$	5339.58	0.45	$B_7 = (5339.58, 0.45)/v_7$
5431.56	0.46	$A_8 = (5431.56, 0.46)/u_8$	5442.57	0.39	$B_8 = (5442.57.0.39)/v_8$
5502.90	0.50	$A_9 = (5502.90, 0.50)/u_9$	5549.68	0.20	$B_9 = (5549.68, 0.20)/v_9$

(Table 22) contd.....

TAIEX Mean	TAIEX STD	Fuzzy Sets of First Factor A_i	TAIFEX Mean	TAIFEX STD	Fuzzy Sets of Second factor B_i
5596.47	0.61	$A_{10} = (5596.47, 0.61)/u_{10}$	5623.24	0.24	$B_{10} = (5623.24, 0.24)/v_{10}$
5674.17	0.80	$A_{11} = (5674.17, 0.80)/u_{11}$	5766.94	0.78	$B_{11} = (5766.94, 0.78)/v_{11}$
5755.49	0.29	$A_{12} = (5755.49, 0.29)/u_{12}$	5919.89	0.30	$B_{12} = (5919.89.0.30)/v_{12}$
5793.94	0.32	$A_{13} = (5793.94, 0.32)/u_{13}$	5995.92	0.33	$B_{13} = (5995.92, 0.33)/v_{13}$
5899.96	0.81	$A_{14} = (5899.96, 0.81)/u_{14}$	6068.37	0.29	$B_{14} = (6068.37, 0.29)/v_{14}$
5950.05	0.28	$A_{15} = (5950.05, 0.28)/u_{15}$	6143.23	0.20	$B_{15} = (6143.23.0.20)/v_{15}$
6092.26	0.35	$A_{16} = (6092.26, 0.35)/u_{16}$	6191.89	0.59	$B_{16} = (6191.89.0.59)/v_{16}$

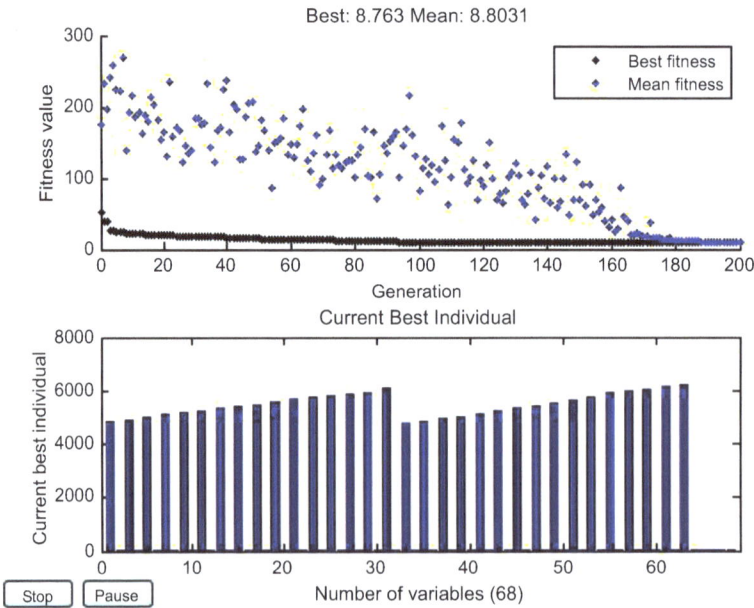

Fig. (9). Best and mean RMSE of 200 generations and best variables of the last generation.

The tuned parameters of operators and defuzzification are as follows:

a. Negation parameter ($p = 0.58$).
b. T-Norm parameter ($q = 0.49$).
c. S-Norm parameter ($r = 0.46$).
d. Defuzzification parameter ($s = 1.50$).

Then aggregated rules with parametric defuzzification should be defuzzified in order to calculate forecasting data. In this example, upper and lower bounds of aggregated rules are defuzzified separately, so the defuzzified or the output of NSIT2 FTS system is:

$$y \text{ (forecasted value)} = \frac{y_U + y_L}{2}$$

Results of the evaluation of the proposed NSIT2 FTS system with tuned parameters has RMSE=8.76 and AFER=0.11% which shows how this system improves the results with a higher degree of accuracy than NS type-1 FTS system. Results are shown in Table **23**.

Also, the tuned parameters are used for NS type-1 FTS system with defuzzification parameter, $s = 1.85$. As shown in Table **23**, columns 5 and 6, the NS type-1 FTS system has better performance with these parameters.

Table 23. Comparison of Forecasted values that obtained by NStype-1 and NS intervaltype-2 FTSs based on tuned parameters and Haurng (2005) method.

Date	Actual TAIEX	Forecasted TAIEX By IT2FTSs	Difference	Forecasted TAIEX By T1FTSs	Difference	Huarng (2005) Method	Difference
11/04/2000	5677.30	5673.23	4.07	5673.30	4.00	5725.00	-47.70
11/06/2000	5657.48	5656.81	0.67	5655.66	1.82	5700.00	-42.52
11/07/2000	5877.77	5876.55	1.22	5880.09	-2.32	5662.50	215.27
11/08/2000	6067.94	6078.45	-10.51	6083.65	-15.71	5750.00	317.94
11/09/2000	6089.55	6086.83	2.72	6089.73	-0.18	6025.00	64.55
11/10/2000	6088.74	6087.71	1.03	6090.66	-1.92	5960.00	128.74
11/13/2000	5793.52	5794.91	-1.39	5794.50	-0.98	6010.00	-216.48
11/14/2000	5772.51	5754.97	17.54	5755.31	17.20	5700.00	72.51
11/15/2000	5737.02	5755.18	-18.16	5755.56	-18.54	5700.00	37.02
11/16/2000	5454.13	5455.93	-1.80	5451.47	2.66	5758.34	-304.21
11/17/2000	5351.36	5356.23	-4.87	5355.56	-4.20	5433.34	-81.98
11/18/2000	5167.35	5169.65	-2.30	5168.39	-1.04	5350.00	-182.65
11/20/2000	4845.21	4845.09	0.12	4842.76	2.45	5150.00	-304.79
11/21/2000	5103.00	5121.74	-18.74	5121.66	-18.66	5150.00	-47.00
11/22/2000	5130.61	5128.58	2.03	5126.93	3.68	5150.00	-19.39
11/23/2000	5146.92	5127.75	19.17	5126.22	20.70	5150.00	-3.08
11/24/2000	5419.99	5422.60	-2.61	5423.27	-3.28	5150.00	269.99
11/27/2000	5433.78	5430.65	3.13	5431.22	2.56	5375.00	58.78

(Table 23) contd.....

Date	Actual TAIEX	Forecasted TAIEX By IT2FTSs	Difference	Forecasted TAIEX By T1FTSs	Difference	Huarng (2005) Method	Difference
11/28/2000	5362.26	5354.67	7.59	5355.04	7.22	5433.34	-71.08
11/29/2000	5319.46	5317.55	1.91	5302.08	17.38	5325.00	-5.54
11/30/2000	5256.93	5258.68	-1.75	5249.45	7.48	5350.00	-93.07
RMSE		**8.76**		**10.16**		**160.943**	
AFER		**0.11%**		**0.13%**		**2.232%**	

Figs. (**10** and **11**) show the comparison between the actual and forecasted daily TAIEX of November year 2000, by proposed NSIT-2 and NS Type-1 FTS systems with genetic algorithm respectively.

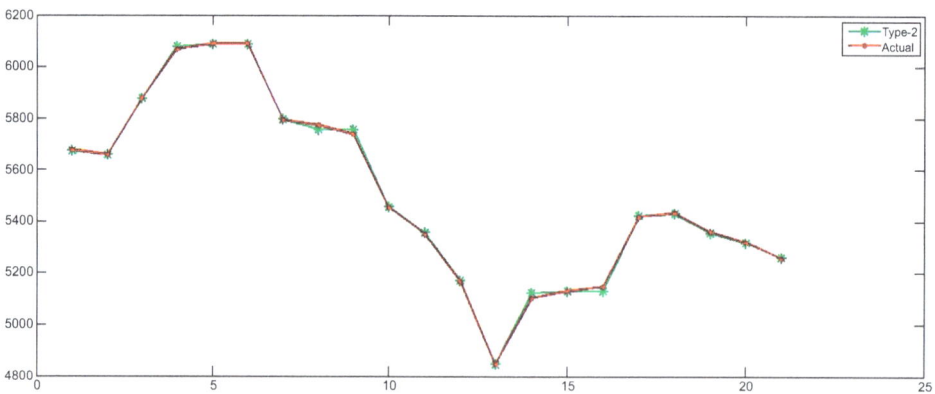

Fig. (10). Comparison between actual and forecasted and daily TAIEX of November year 2000 which obtained by proposed NSIT-2 FTS system with genetic algorithm.

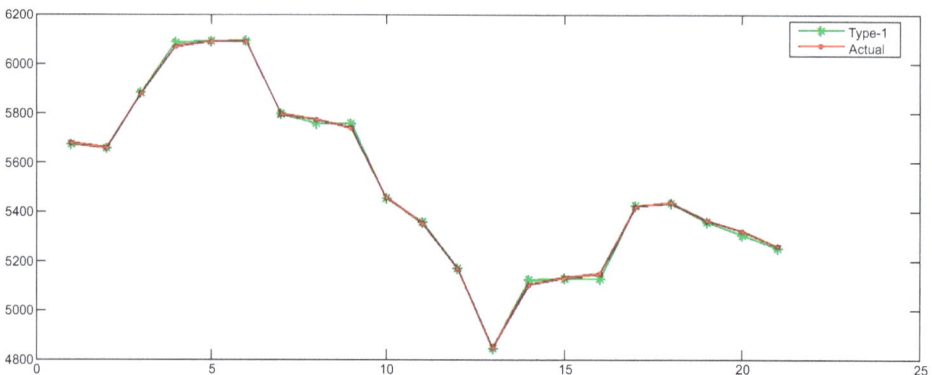

Fig. (11). Comparison between the Actual and forecasted and daily TAIEX of November year 2000 which obtained by proposed NS Type-1 FTS with genetic algorithm.

CONCLUSION AND FUTURE WORKS

This paper presents a general concept for fuzzy time series to resolve the drawbacks of previous methods, which were fuzzy time series systems with crisp inference. In this paper, we have generated a rule based two factors high order non singleton type-1 and interval type-2 fuzzy time series forecasting system. This system consists of type-1 inputs, Sharp Gaussian membership function, third order two factors fuzzy rules, Mamdani inference engine with FITA approach and BADD parametric defuzzification, and RMSE and AFER model validation. The proposed forecasting method is applied to forecast daily average temperature from June 1, 1996 to September 30, 1996 in Taipei and daily TAIEX index for November 2000. This system is capable of predicting the second factor easily by replacing consequents of rules with it. In addition, a genetic algorithm is applied to tune the main parameters of the interval type-2 fuzzy system. The results show that the proposed method is more accurate and reliable than all previous ones. In the second example, it is shown how type-2 FLSs have higher degree of accuracy than type-1 FLSs. Moreover, in this research it is shown how length and number of intervals, number of orders, type of membership functions, type of operators, the type reduction and defuzzification methods and the domain applied for the inference area would affects forecasting systems and forecasted results and accuracy rate of the system.

For future studies general type-2 fuzzy logic systems could be applied instead of interval ones to represent more accurate methods.

CONFLICT OF INTEREST

The authors (editor) declares no conflict of interest, financial or otherwise.

ACKNOWLEDGEMENTS

Declared none.

REFERENCES

[1]   S.M. Chen, "Forecasting enrolments based on fuzzy time series", *Fuzzy Sets Syst.,* vol. 81, pp. 311-319, 1996.
[http://dx.doi.org/10.1016/0165-0114(95)00220-0]

[2]   S.M. Chen, and J.R. Hwang, "Temperature prediction using fuzzy time series", *IEEE Trans. Syst. Man Cybern. B Cybern.,* vol. 30, no. 2, pp. 263-275, 2000.
[http://dx.doi.org/10.1109/3477.836375]

[3]   L. Davis, *Handbook of genetic algorithms.* Van Nostra and Reinhold, 1991. ISBN: 04- 4200-1783.

[4]   L.Y. Hsu, "Temperature prediction and TAIFEX forecasting based on fuzzy relationships and MTPSO techniques", *Expert Syst. Appl.,* vol. 37, pp. 2756-2770, 2010.
[http://dx.doi.org/10.1016/j.eswa.2009.09.015]

[5]   J. Holland, *Adaptation in natural and artificial systems: An introductory analysis with applications to Biology, control and artificial intelligence.* MIT Press: London, 1992.

[6]   K. Huarng, and H.K. Yu, "Type-2 Time Series Model for Stock Index Forecasting", *Physica A,* vol. 353, pp. 445-462, 2005.
[http://dx.doi.org/10.1016/j.physa.2004.11.070]

[7]   N.N. Karnik, J.M. Mendel, and Q. Liang, "Type-2 fuzzy logic systems", *IEEE Trans. Fuzzy Syst.,* vol. 7, no. 6, pp. 643-658, 1999.
[http://dx.doi.org/10.1109/91.811231]

[8]   N.N. Karnik, and J.M. Mendel, "Type-2 fuzzy logic systems: type-reduction", *IEEE Conference on Systems, Man and Cybernetics,* 1998 La Jolla, CA.

[9]   L.W. Lee, L.H. Wang, S.M. Chen, and Y.H. Leu, "Handling forecasting problems based on two-factor high-order fuzzy time series", *IEEE Trans. Fuzzy Syst.,* vol. 14, pp. 468-477, 2006.
[http://dx.doi.org/10.1109/TFUZZ.2006.876367]

[10]   L.W. Lee, L.H. Wang, and S.M. Chen, "Temperature prediction and TAIFEX forecasting based on fuzzy logical relationships and genetic algorithms", *Expert Syst. Appl.,* vol. 33, pp. 539-550, 2007.
[http://dx.doi.org/10.1016/j.eswa.2006.05.015]

[11]   L.W. Lee, L.H. Wang, and S.M. Chen, "Temperature prediction and TAIFEX forecasting based on high-order fuzzy logical relationships and genetic simulated annealing techniques", *Expert Syst. Appl.,* vol. 34, pp. 328-336, 2008.
[http://dx.doi.org/10.1016/j.eswa.2006.09.007]

[12]   Z. Li, Z. Chen, and J. Li, "A model of weather forecast by fuzzy grade statistics", *Fuzzy Sets Syst.,* vol. 26, pp. 275-281, 1988.
[http://dx.doi.org/10.1016/0165-0114(88)90123-6]

[13]   Q. Liang, and J.M. Mendel, "Interval type-2 fuzzy logic systems: Theory and design", *IEEE Trans. Fuzzy Syst.,* vol. 8, pp. 535-550, 2000.
[http://dx.doi.org/10.1109/91.873577]

[14]   J.M. Mendel, *Uncertain rule-based fuzzy logic systems: introduction and new directions.* Prentice-Hall: Upper-Saddle River, NJ, 2001.

[15]   J.M. Mendel, R.I. John, and F. Liu, "Interval Type-2 fuzzy logic systems made simple", *IEEE Trans. Fuzzy Syst.,* vol. 14, no. 6, pp. 808-821, 2006.
[http://dx.doi.org/10.1109/TFUZZ.2006.879986]

[16]   J.M. Mendel, and R.I. John, "Type-2 fuzzy sets made simple", *IEEE Trans. Fuzzy Syst.,* vol. 10, no. 2, pp. 117-127, 2002.
[http://dx.doi.org/10.1109/91.995115]

[17]   J.M. Mendel, and H. Wu, "Type-2 fuzzistics for symmetric interval type-2 fuzzy sets: Part 1, forward problems", *IEEE Trans. Fuzzy Syst.,* vol. 14, no. 6, 2006.
[http://dx.doi.org/10.1109/TFUZZ.2006.881441]

[18]   J.M. Mendel, and H. Wu, "Type-2 fuzzistics for symmetric interval type-2 fuzzy sets: Part 1, inverse problems", *IEEE Trans. Fuzzy Syst.,* vol. 15, no. 2, 2007.
[http://dx.doi.org/10.1109/TFUZZ.2006.881447]

[19]   Q. Song, and B.S. Chissom, "Fuzzy time series and its models", *Fuzzy Sets Syst.,* vol. 54, pp. 269-277, 1993.
[http://dx.doi.org/10.1016/0165-0114(93)90372-O]

[20]   Q. Song, and B.S. Chissom, "Forecasting enrollments with fuzzy time series– Part I", *Fuzzy Sets Syst.,* vol. 54, pp. 1-9, 1993.
[http://dx.doi.org/10.1016/0165-0114(93)90355-L]

[21]   Q. Song, and B.S. Chissom, "Forecasting enrollments with fuzzy time series – Part II", *Fuzzy Sets*

Syst., vol. 62, pp. 1-8, 1994.
[http://dx.doi.org/10.1016/0165-0114(94)90067-1]

[22]   I.B. Turksen, "Fuzzy Normal Forms", *Fuzzy Sets Syst.,* pp. 253-266, 1994.

[23]   I.B. Turksen, *An Ontological and Epistemological Perspective of Fuzzy Set Theory.* Elsevier Inc., 2006, pp. 27-30.

[24]   N.Y. Wang, and S.M. Chen, "Temperature prediction and TAIFEX forecasting based on automatic clustering techniques and two-factor high-order fuzzy time series", *Expert Syst. Appl.,* vol. 36, pp. 2143-2154, 2009.
[http://dx.doi.org/10.1016/j.eswa.2007.12.013]

[25]   D. Wu, and W.W. Tan, "A type-2 fuzzy logic controller for the liquid-level process", *Proceedings of IEEE international conference on fuzzy systems Budapest,* 2004, pp. 953-958, Hungary.
[http://dx.doi.org/10.1109/FUZZY.2004.1375536]

[26]   R.R. Yager, and D.P. Filev, *Essentials of fuzzy modelling and control.* Wiley: New York, 1994.

[27]   R.R. Yager, "On a general class of fuzzy connectives", *Fuzzy Sets Syst.,* vol. 4, no. 3, pp. 235-242, 1980.
[http://dx.doi.org/10.1016/0165-0114(80)90013-5]

[28]   L.A. Zadeh, "Fuzzy sets", *Inf. Control,* vol. 8, pp. 338-353, 1965.
[http://dx.doi.org/10.1016/S0019-9958(65)90241-X]

[29]   L.A. Zadeh, "The concept of a linguistic variable and its application to approximate reasoning", *Inf. Sci.,* vol. 8, pp. 199-249, 1975.
[http://dx.doi.org/10.1016/0020-0255(75)90036-5]

CHAPTER 4

A New Neural Network Model with Deterministic Trend and Seasonality Components for Time Series Forecasting

Erol Egrioglu[1,*], Cagdas Hakan Aladag[2], Ufuk Yolcu[3], Eren Bas[1] and **Ali Z. Dalar[1]**

[1] *Department of Statistics, Faculty of Arts and Sciences, Giresun University, Giresun, Turkey*

[2] *Department of Mechanical and Industrial Engineering, University of Toronto, Toronto, Canada*

[3] *Department of Statistics, Faculty of Science, Ankara University, Ankara, Turkey*

Abstract: Artificial neural networks have been commonly used for time series forecasting problem in the last years. When they are compared with classical time series methods, artificial neural networks have some advantages. Artificial neural networks do not need any assumption such as normality and linearity. In recent years, different types of artificial neural networks have been proposed for time series forecasting. In these networks, the inputs are lagged variables or other time series. It is well known that some time series have deterministic trend and this kind of time series should be modeled by using different functions of time (t) as inputs. In the modeling such type time series, using only lagged variables will lead to insufficient results. In this study, a new neural network model that has different functions of time as inputs is proposed for solving this problem. The proposed method is compared with other methods in the literature according to forecast performance. It is obtained that the new model outperforms other methods.

Keywords: Artificial neural networks, Forecasting, Particle swarm optimization, Seasonality, Time series.

INTRODUCTION

Artificial neural networks can produce accurate forecasts for real world time series. When artificial neural networks have been employed for time series forecasting, lagged variables of original time series have been used as inputs of network. In classical time series analysis, many kinds of probabilistic models were employed. These methods can be classified into 2 classes. The inputs of first

* **Corresponding author Erol Egrioglu:** Giresun University, Faculty of Arts and Sciences, Department of Statistics, 28100, Giresun, Turkey; E-mail: erole1977@yahoo.com

type models are lagged variables whereas the inputs of second type model are functions of time (t). Two type of models are given in following equations.

$$X_t = f(X_{t-1}, X_{t-2}, \dots, X_{t-k})$$ **(1)**

$$X_t = f(t)$$ **(2)**

In two equations, f function can be linear or nonlinear function. f function can be selected to provide for modelling seasonality. If a time series have both of trend and seasonality, suitable model should be used for this kind of time series. In the literature, deterministic functions of time (t) were not employed in neural networks. There is only one study Teixeira and Fernandes [1] that employs time index as input in multilayer perceptron neural network. Teixeira and Fernandes [1] did not use deterministic trend and seasonality functions as inputs. In this study, a new artificial neural network model was proposed. The new neural network model is a kind of combined model. It has four components. These are liner autoregressive part, nonlinear autoregressive part, deterministic trend and deterministic seasonality part. The new neural network model can use for modelling times series which have trend and seasonality. The review of classical time series models is given in second section. In the third section, artificial neural networks are summarized for time series forecasting problem. The proposed neural network is introduced in the fourth section and the application results are given in the fifth section. Final section is about the conclusions of the study.

CLASSICAL TIME SERIES FORECASTING MODELS

Many forecasting methods are used to obtain forecasts for real world time series. In the classical time series analysis, the model is preferred according to structure of time series. Classical time series models have some assumptions about time series. These assumptions can be listed as deterministic trend, additive or multiplicative trend and seasonality, normality, linearity. If it is assumed that time series have deterministic trend, some regression models can be used. Some of these regression models are given below:

$$X_t = a + bt + \varepsilon_t \qquad \text{(Linear Model)} \qquad \textbf{(3)}$$

$$X_t = a + bt + ct^2 + \varepsilon_t \qquad \text{(Quadratic Model)} \qquad \textbf{(4)}$$

$$X_t = a + bt + ct^2 + dt^3 + \varepsilon_t \qquad \text{(Cubic Model)} \qquad \textbf{(5)}$$

$$X_t = ab^t \varepsilon_t \qquad \text{(Compound Model)} \qquad \textbf{(6)}$$

$$X_t = e^{a+bt} \varepsilon_t \qquad \text{(Growth Model)} \qquad \textbf{(7)}$$

$$X_t = a + b\ln(t) + \varepsilon_t \qquad \text{(Logarithmic Model)} \qquad \textbf{(8)}$$

$$X_t = e^{a+b/t} \varepsilon_t \qquad \text{(S-shaped Model)} \qquad \textbf{(9)}$$

$$X_t = ae^{bt} \varepsilon_t \qquad \text{(Exponential Model)} \qquad \textbf{(10)}$$

$$X_t = a + \frac{b}{t} + \varepsilon_t \qquad \text{(Inverse Model)} \qquad \textbf{(11)}$$

$$X_t = at^b \varepsilon_t \qquad \text{(Power Model)} \qquad \textbf{(12)}$$

$$X_t = 1/\left(\left(\frac{1}{u}\right) + ab^t \varepsilon_t\right) \qquad \text{(Logistic Model)} \qquad \textbf{(13)}$$

where a, b, c, and d are model parameters, ε_t is model error term. These models also called as deterministic trend models. These models have different properties. If a model is a linear form of its parameters, this model is called linear models. (3-5), (8) and (11) models are linear models and they can be estimated by using linear least square method. But the others can be estimated by using linear least square method after applying proper transformation. (6), (7), (9), (10), and (12) models can be linearized by using natural logarithmic transformation. (13) model can be linearized by using two consecutive transformations. If residuals is not white noise series after applying (3-13) models to time series, time series has different properties as well as deterministic trend and it is needed to use different models like autoregressive integrated moving average (ARIMA).

If time series has trend and seasonality, the model contains deterministic trend terms and seasonal terms for the seasons. Before using seasonal deterministic trend models, period of time series is determined. Seasonal deterministic trend models have a prior assumption about structure of trend and seasonality. Two structures were used in these models. These structure can be given in following equations.

$$X_t = T_t + S_t + \varepsilon_t \qquad \textbf{(14)}$$

$$X_t = T_t \times S_t + \varepsilon_t \qquad \textbf{(15)}$$

$$\hat{L}_T = \alpha_1 \left(\frac{X_T}{\hat{s}_{T-s}}\right) + (1 - \alpha_1)(\hat{L}_{T-1} + \hat{\beta}_{T-1}) \tag{16}$$

$$\hat{\beta}_T = \alpha_2(\hat{L}_T - \hat{L}_{T-1}) + (1 - \alpha_2)\hat{\beta}_{T-1} \tag{17}$$

$$\hat{s}_T = \alpha_3 \left(\frac{X_T}{\hat{L}_T}\right) + (1 - \alpha_3)\hat{s}_{T-s} \tag{18}$$

$$\hat{X}_{T+1} = (\hat{L}_T + \hat{\beta}_T) \times \hat{s}_{T-s} \tag{19}$$

Similar formulas are used for AWES, the details of AWES and MWES can be reviewed from Makridakis *et al.* [2].

If time series have trend and seasonal components, other alternative forecasting methods are regression methods with harmonics or dummy variables for seasons. Regression methods with harmonics have additive or multiplicative structures of trend and seasonality. Additive regression model with harmonics is given below.

$$X_t = \beta_0 + \sum_{i=1}^{m} \beta_i t^i + \sum_{j=1}^{[\![s/2]\!]} \left[\alpha_j \sin\left(\frac{2\pi jt}{s}\right) + \gamma_j \cos\left(\frac{2\pi jt}{s}\right)\right] \tag{20}$$

The last term of the equation contains [s/2] harmonics. The first term is used for modelling trend of time series. β_i, $i = 1,2,\ldots, m$, α_j and γ_j, $j = 1,2,\ldots, [s/2]$ are parameters of the regression model. Model can be easily modified for multiplicative structure of trend and seasonal terms. Another regression methods can be constructed by using dummy variables. The number of dummy variables must be number of seasons minus one because of linear dependence problem of design matrix columns. The regression methods with dummy variables can be construct in additive or multiplicative structures, too. The additive regression model with dummy variables is given in equation (21).

$$X_t = \beta_0 + \sum_{i=1}^{m} \beta_i t^i + \gamma_1 D_1 + \cdots + \gamma_{s-1} D_{s-1} + \varepsilon_t \tag{21}$$

In (21), β_i, $i = 1,2,\ldots, m$ and γ_j, $j = 1,2,\ldots,s-1$ are parameters of regression model. It is clear that the model is a linear model. The parameters of the model can be estimated by using linear least square method. If residuals is not white noise series after applying seasonal methods to time series, time series has different properties as well as deterministic trend and seasonality, it is needed to use different models like seasonal autoregressive integrated moving average (SARIMA).

ARIMA and SARIMA methods can be used for forecasting time series with trend or seasonal components, respectively. ARIMA and SARIMA models are given equations (22) and (23).

$$\phi(B)(1 - B)^d (X_t - \mu) = \theta(B)\varepsilon_t \tag{22}$$

$$\phi(B)\Phi(B)(1 - B)^d (1 - B^s)^D (X_t - \mu) = \theta(B)\Theta(B)\varepsilon_t \tag{23}$$

where s is period of time series and μ is mean of time series and polynomials are given following equations.

$$\phi(B) = 1 - \phi_1 B - \cdots - \phi_p B^p \tag{24}$$

$$\theta(B) = 1 - \theta_1 B - \cdots - \theta_q B^q \tag{25}$$

$$\Phi(B) = 1 - \Phi_1 B^s - \cdots - \Phi_P B^{P \times s} \tag{26}$$

$$\Theta(B) = 1 - \Theta_1 B - \cdots - \Theta_Q B^{Q \times s} \tag{27}$$

In these models, stationarity and invertibility are two critical assumptions. To obtain stationary time series, proper value of difference parameters *dandD* must be determined. In the literature, source of non-stationarity was discovered by using unit root tests. According to result of unit root tests, de-trending or differencing methods are applied to obtain stationary time series. Moreover, stationarity and invertibility conditions are controlled in parameter estimation processes of models. ARIMA and SARIMA methods are not linear models when $q \neq 0$ or $Q \neq 0$. As a result of this, non-linear least square methods must be used to obtain parameter estimations of these models. Levenberg Marquardt (LM) method for these models was used in Box and Jenkins [3]. LM method is based on second order derivatives of the objective function. An algorithm based on first order derivatives was given in Wei [4].

ARTIFICIAL NEURAL NETWORKS FOR FORECASTING TIME SERIES

Artificial neural networks imitate human neural system and it is a very simple model of human neural system. Human neural system is very complicated and it is well known that the system works with three billions neurons. Artificial neuron models are very simple and they are very bad sample of real biological neuron model. Although artificial neuron models are not exact model of biological neuron, the network of artificial neuron models with a few neurons can give successful prediction results for real world data. It is not a miracle, the artificial neural networks are nonlinear and databased models. If they are used correctly, it is possible to obtain better results than classical inference methods like regression methods. In recent years, many of artificial neural network models have been proposed in the literature. The most preferred artificial neural networks are multilayer perceptron (MLP) models. To obtain forecasts from time series, inputs of MLP are lagged variables and output of MLP is original time series. A MLP architecture for forecasting is given in Fig. (**1**).

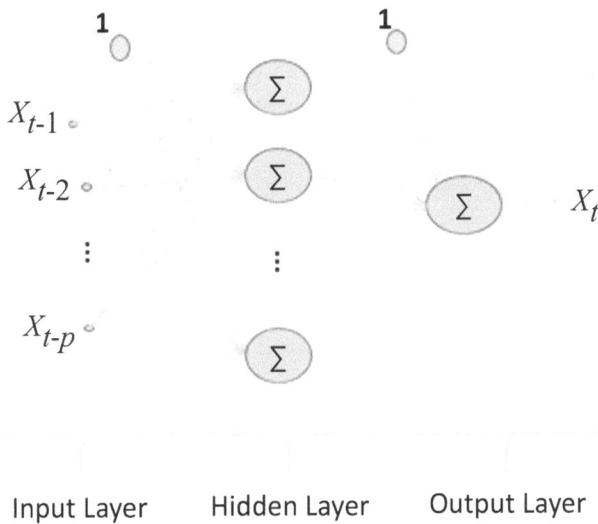

Fig. (1). Multilayer perceptron artificial neural network with one output neuron.

MLP is a feed forward neural network, its weights and biases are its parameters. MLP presents a nonlinear form of these parameters. The parameters of MLP can be estimated by using nonlinear least square methods. Backpropagation (BP) algorithm was proposed for MLP parameter estimating in Rumelhart *et al.* [5]. The many of modified versions of BP algorithm have been proposed in the literature. Recent years, artificial intelligence optimization techniques have been used to estimate parameters of MLP. Some of these techniques are particle swarm optimization, genetic algorithm and artificial bee colony. Artificial intelligent optimization techniques do not need to compute derivatives and they have smaller probability to trap local optimum than derivate based algorithms like BP and LM.

Elman artificial neural networks (E-ANN) have been used to obtain forecasts of time series in the literature. E-ANN was proposed by Elman [6]. E-ANN is a recurrent neural network, its outputs are also its inputs of hidden layer context units. The architecture of E-ANN is given in Fig. (2).

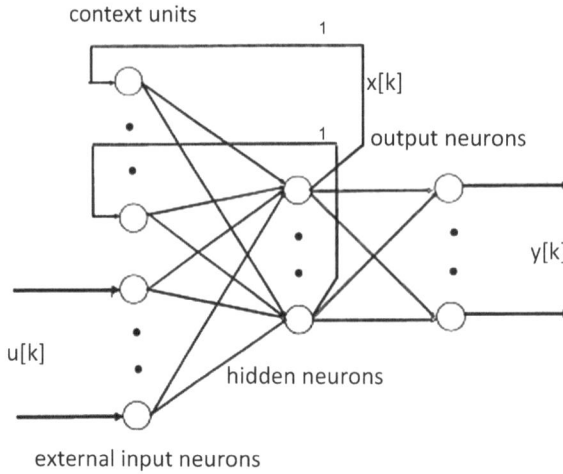

Fig. (2). Elman artificial neural network with k output neurons.

If the weights of context units' connections are assumed as fixed, BP algorithm can be employed for estimation parameters of E-ANN. Artificial intelligent optimization algorithms can be used without any restrictions on the weights for estimating parameters of E-ANN.

Another important artificial neural network type is single multiplicative neuron model artificial neural network (SMNM-ANN). This neural network was firstly proposed by Yadav *et al.* [7]. SMNM-ANN has only a neuron. Because of this, SMNM-ANN has less parameters than MLP-ANN or E-ANN. The aggregation function of the neuron is multiplicative. It was shown that SMNM-ANN can produce good prediction results for real world data sets in [7]. In Fig. (3), the architecture of SMNM-ANN is shown.

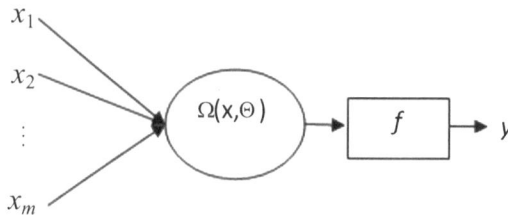

Fig. (3). Single multiplicative neuron model artificial neural network.

In Yadav *et al.* [7], backpropagation algorithm was given as a training algorithm of the SMNM-ANN. Zhao and Yang [8] and Samanta [9] used particle swarm optimization method as a training algorithm of SMNM-ANN. In the literature, many artificial neural network types can be proposed based on multiplicative neuron model. Linear and nonlinear neural network (L&NL-ANN) is one of them. L&NL-ANN employs multiplicative and additive neuron models. L&NL-ANN was proposed by Yolcu *et al.* [10]. The architecture of the L&NL-ANN is given in Fig. (**4**).

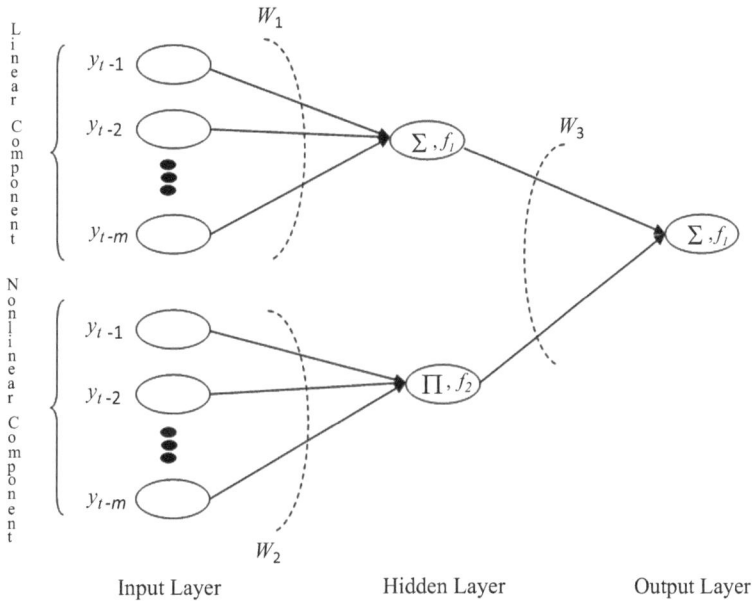

Fig. (4). Linear and nonlinear artificial neural network.

L&NL-ANN has three components. The first component is linear component and the linear aggregation function and linear activation function were used in this component. The second component is non-linear component and multiplicative aggregation function and logistic activation function were used in this component. The third component is combination component and linear aggregation and activation functions were employed in this component. Yolcu *et al.* [10] proposed to use modified particle swarm optimization for training L&NL-ANN.

Multiplicative seasonal artificial neural network (MS-ANN) was proposed by Aladag *et al.* [11]. MS-ANN has three components as trend and seasonality. The third component provides multiplication of trend and seasonal components. In MS-ANN, additive and multiplicative neuron models were employed. MS-ANN is a kind of nonlinear multiplicative seasonal time series forecasting model.

Aladag *et al.* [11] used modified particle swarm optimization algorithm to train MS-ANN. The architecture of MS-ANN is given in Fig. (**5**).

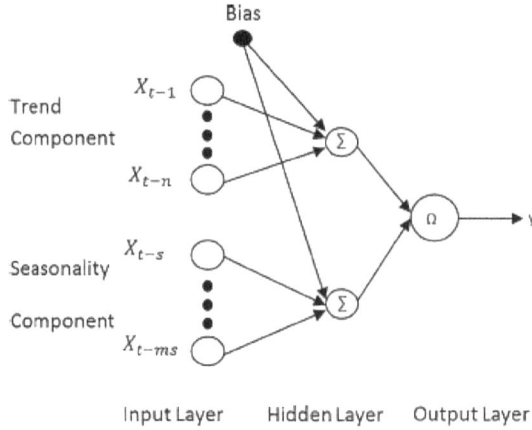

Fig. (5). Multiplicative seasonal artificial neural network.

Recurrent multiplicative neuron model artificial neural network (RMNM-ANN) was proposed by Egrioglu *et al.* [12]. RMNM-ANN is a recurrent version of MNM-ANN. MNM-ANN is a kind of nonlinear autoregressive model. RMNM-ANN is kind of nonlinear autoregressive moving average model. The outputs of RMNM-ANN are also inputs of the network. Training of RMNM-ANN was performed by using modified particle swarm optimization in Egrioglu *et al.* [12]. The architecture of RMNM-ANN is given in Fig. (**6**).

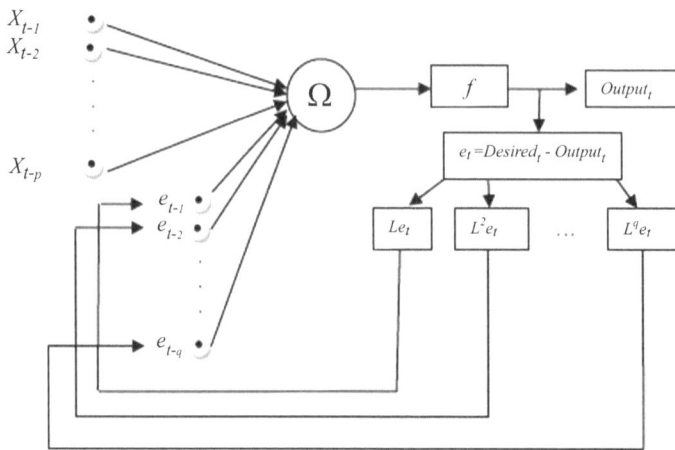

Fig. (6). Recurrent multiplicative neuron model artificial neural network.

A NEW ARTIFICIAL NEURAL NETWORK WITH DETERMINISTIC COMPONENTS

In this study, a new artificial neural network is proposed for forecasting purpose. New neural network is call deterministic components artificial neural network (DC-ANN). DC-ANN is a hybrid network of regression models, linear autoregressive model and MNM-ANN. DC-ANN has four type inputs. The first type inputs are lagged variables of normalized time series, the second type is lagged variables of time series, third type is powers of time and fourth type is dummy variables for seasons. DC-ANN is suitable for time series which has trend and seasonality. DC-ANN is a first neural network model that is combine classical regression models and artificial neural network. DC-ANN is an artificial network, it is not a classical two staged hybrid method. The architecture of DC-ANN is given in Fig. (**7**).

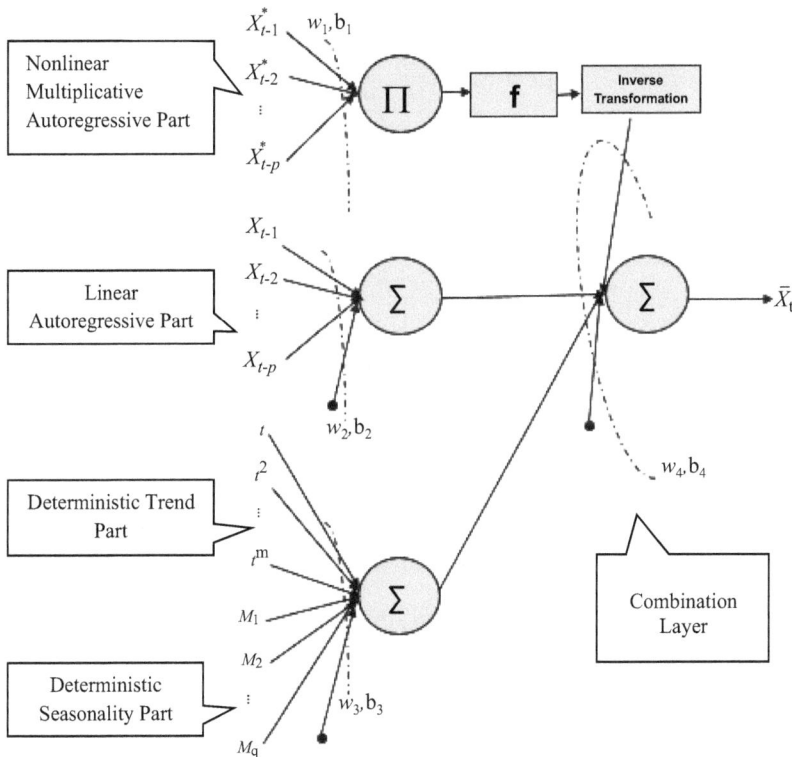

Fig. (7). The architecture of DC-ANN.

DC-ANN is explained by two algorithms. The first algorithm gives details of calculation an output from DC-ANN for a learning sample. The second algorithm gives details of training method. DC-ANN is trained by using guaranteed convergence particle swarm optimization that was proposed in Van Der Bergh and Engelbretch [13].

Algorithm 1. Calculation of an output of DC-ANN for a learning sample

Step 1. Constitute a learning sample by using transformed and original time series data as below:

$$\left[x_{t-1}^*, \dots x_{t-p}^* \; ; x_{t-1}, \dots, x_{t-p}; t, t^2, \dots, t^m; \sin\left(\frac{2\pi t}{s}\right), \cos\left(\frac{2\pi t}{s}\right), \dots, \right.$$
$$\left. \sin\left(\frac{2\pi t[\![s/2]\!]}{s}\right), \cos\left(\frac{2\pi t[\![s/2]\!]}{s}\right) \right]$$

where x_t is original time series observation, x_t^* is transformed time series observation, t is time index and it is 1 for the first observation and s is period of time series. The transformation is applied by using (28) equation.

$$x_t^* = \frac{x_t - \min(x_t)}{\max(x_t) - \min(x_t)} \tag{28}$$

Step 2. Calculate output of nonlinear multiplicative autoregressive part by using equation (29), (30) and (31).

$$net1 = \prod_{i=1}^{p}(w_1(i)x_{t-i}^* + b_1(i)) \tag{29}$$

$$o_1^* = \frac{1}{1 + \exp(-net1)} \tag{30}$$

$$o_1 = (\max(x_t) - \min(x_t)) \times o_1^* + \min(x_t) \tag{31}$$

Step 3. Calculate output of linear autoregressive part by using equation (32) and (33).

$$net2 = b_2 + \sum_{i=1}^{p} w_2(i)\, x_{t-i} \tag{32}$$

$$o_2 = net2 \tag{33}$$

Step 4. Calculate output of deterministic trend and seasonality part by using equation (34) and (35).

$$net3 = b_3 + w_3(1)t + \cdots + w_3(m)t^m + w_3(m+1)\sin\left(\frac{2\pi t}{s}\right) +$$

$$w_3(m+2)\cos\left(\frac{2\pi t}{s}\right) + \cdots + w_3(m+q-1)\sin(\frac{\pi tq}{s}) + w_3(m+q)\sin(\frac{\pi tq}{s}) \quad \textbf{(34)}$$

$$o_3 = net3 \quad \textbf{(35)}$$

Step 5. Calculate output of the network by using equation (36).

$$\hat{x}_t = b_4 + w_4(1)o_1 + w_4(2)o_2 + w_4(3)o_3 \quad \textbf{(36)}$$

Algorithm 2. Training of DC-ANN by using particle swarm optimization

Step 1: Initial vectors including velocities and positions of the particles are randomly constructed. Positions of the particles are the parameters of weights of DC-ANN. In order to exemplify, an example is presented in Fig. (8) with positions of a particle with $p = 2$, $m = 2$, $q = 2$ in DC-ANN. All initial positions of the particle are randomly obtained from interval (0, 1). However, velocities are randomly generated from interval *(-vmaps,vmaps)* depending on the pre-determined *vmaps* limit value.

P1	P2	P3	P4	P5	P6	P7	P8
$w_1(1)$	$w_1(2)$	$b_1(1)$	$b_1(2)$	$w_2(1)$	$w_2(2)$	b_2	$w_3(1)$
P9	P10	P11	P12	P13	P14	P15	P16
$w_3(2)$	$w_3(3)$	$w_3(4)$	b_3	$w_4(1)$	$w_4(2)$	$w_4(3)$	b_4

Fig. (8). Positions of a particle in DC-ANN when $p = 2$, $m = 2$, $q = 2$.

Step 2: Fitness function values for all particles are computed. Root mean square error (RMSE) value, whose formula is given below, is obtained by using outputs obtained from DC-ANN for learning samples as fitness function.

$$RMSE = \sqrt{\frac{1}{n}\sum_{t=1}^{n}(y_t - \hat{y}_t)^2} \quad \textbf{(37)}$$

where $\hat{y}_t$ is the output of DC-ANN, y_t is the target value and n is the number of training examples. In order to calculate RMSE value for each particle, outputs of DC-ANN $\hat{y}_t, t = 1,2,\cdots,n$ are calculated by using the values of positions of the particle and Algorithm 1.

Step 3: *Pbest* and *Gbest* are updated. If the maximum number of iterations is reached, or the fitness value of *Gbest* is less than a predetermined value of ε then, the process is determined. Otherwise, go to Step 4.

Step 4: Update positions and the velocity values of the positions then, go back to Step 2. Update operations are performed by using the formulas given in equations (38)-(41).

$$v_{ij}^{k+1} = w \times v_{ij}^k + c_1 r_1 (pbest_{ij} - x_{ij}) + c_2 r_2 (gbest_j - x_{ij}) \qquad \textbf{(38)}$$

$$x_{ij}^{k+1} = x_{ij}^k + v_{ij}^{k+1} \qquad \textbf{(39)}$$

where w is the inertia weight, c_1 is the cognitive coefficient and c_2 is the social coefficient. r_1 and r_2 are numbers randomly generated from uniform distribution with (0,1) parameters. $Pbest_{ij}$ is the best condition for particle i^{th} at j^{th} position. $Gbest_i$ is the best condition for all particles at j^{th} position. x_{ij} is the position value of ith particle at j^{th} position. v_{ij} is the jth velocity of ith particle and the related formula is given in (40).

$$v_{ij}^{k+1} = w * v_{ij}^k - x_{ij}^k + gbest_j + \rho(k) * r_3 \qquad \textbf{(40)}$$

where, k is a variable representing the number of iterations. r_3 is randomly generated from (-1,1) uniform distribution. In each iteration, $\rho(k + 1)$ is computed by using the following formula.

$$\rho(k+1) = \begin{cases} 2\rho(k), s_n \rangle s_c \\ 0.5\rho(k), f_n \rangle f_c \\ \rho(k), o.w. \end{cases} \qquad \textbf{(41)}$$

where, $\rho(0) = 1$, s_n is the number of success, f_n is the number of failure and s_c and f_c are the upper limit for success and failure respectively. If an improvement is achieved without changing the best particle's number, f_n is reset (set to zero) and s_n is increased. Otherwise, f_n is increased and s_n is reset. Both the number of failures and successes are reset, if *Gbest* has a different particle number.

APPLICATIONS

In order to show the performance of DC-ANN, two real world time series are employed in the implementation. The first one is beer consumption in Australia which is observed from 1956 to 1994 [14]. The graph of this time series that has both trend and seasonal components is shown in Fig. (**9**). In this graph, the observation values are presented as mega liters in the vertical axis and the horizontal axis represents the time. The data has 148 observations. The last 16 observations were used for testing and the remaining 132 observations were used foe training.

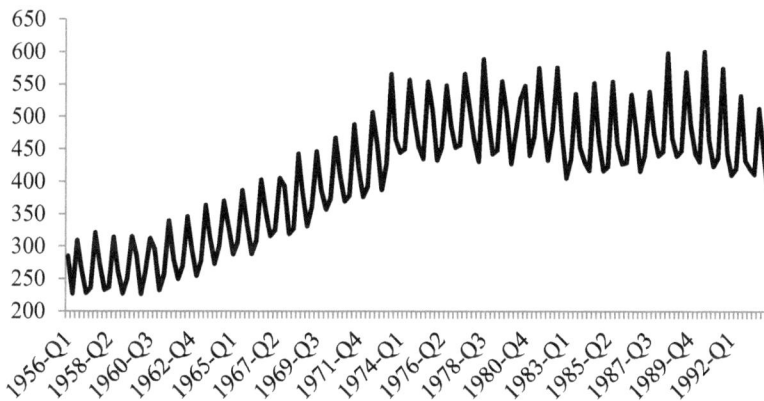

Fig. (9). Beer Consumption in Australia.

The time series was forecasted by utilizing methods such as MLP-ANN, L&NL-ANN, E-ANN, MS-ANN, RMNM-ANN, and DC-ANN. All obtained forecasting results are summarized in Table **1**.

In the application of DC-ANN, parameters of GC-PSO used in the training of RB-MNM-ANN, were taken as follows: *vmaps* = 1, c_1 = 2, c_2 = 2, f_c = 5, s_c = 5, ε = 10^{-6}, the number of particles as 30 and the maximum number of iterations as 1000. Different combinations of p,q and m were tried to reach the best results. The best results were obtained when p = 8, q = 2 and m = 2. The optimal weights of DC-ANN is given in Table **2** when p = 8, q = 2 and m = 2.

Table 1. The performance criteria values for test data of Australian beer consumption data.

	MLP-ANN	L&NL-ANN	E-ANN	MS-ANN	RMNM-ANN	DC-ANN
RMSE	24.1052	18.7888	22.6581	22.1700	17.8573	17.0923
MAPE	0.0476	0.0357	0.0436	0.0393	0.0329	0.0306

Table 2. The optimal weights of DC-ANN when $p = 8$, $q = 2$ and $m = 2$.

	Optimal Weights							
w1	-0.0384	-0.9115	-0.0808	1.3253	0.5591	1.5820	0.1430	-0.0430
w2	-0.3745	-0.2506	2.0070	8.0022	0.3370	0.2544	-1.3356	0.3340
w3	0.2212	0.0021	-0.2240	0.3782	-0.0869			
w4	0.0210	0.1126	-0.2906					
b1	1.2684	0.5412	0.1154	1.0755	-0.1603	1.5247	0.4076	1.4254
b2	0.5059							
b3	1.2891							
b4	0.8673							

The second time series is Turkey Electricity consumption data that is obtained from Turkish Energy Ministry. Monthly data was obtained from January 2002 to December 2013. The graph of time series is given in Fig. (**10**). The last twelve observations are taken as test set and others are training set.

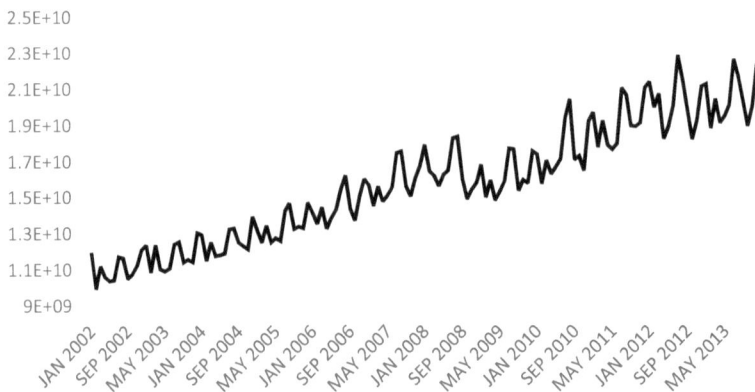

Fig. (10). Turkey Electricity Consumption data.

MLP-ANN, SARIMA, MNM-ANN and L&NL-ANN method were applied to the time series as benchmarks. DC-ANN method was applied to the time series by using parameter selection as first application. The obtained forecasting results are showed in Table **3**. The best results of the proposed method were obtained when p

$= 12$, $q = 2$, and $m = 6$.

Table 3. The performance criteria values and forecasts for test data of Turkey Electricity Consumption Time Series.

Test Data	MLP-ANN	SARIMA $(0,1,1)(0,1,1)_{12}$	MNM-ANN	L&NL-ANN	DC-ANN
21275408487	21504260392	21690314907	21929976335	22005935580	21964102648
18841712637	21407932217	19879748318	19674336447	21273737442	21359220311
20463933683	19934741061	20679491118	20759353666	20386359964	20297900729
19139248871	17339157313	18616749529	18177421217	19068902087	18810598848
19511728912	19132107582	18166751109	19502618441	19337714416	19351318386
20132602347	20521392811	19476718063	19961628854	20493224715	20168250837
22648523194	21559624073	22996373189	21466603203	22493619106	22584712456
21698207982	21608712088	22602364226	20970484983	22642925619	22419574003
20358717408	20450417618	19372692274	20819409088	20387962879	20704520697
18964661109	19608790983	19145368564	18842139533	19384311943	19317372506
20061232838	20798229864	18281990717	20341486661	20668793095	20393976200
22405662577	21396071971	21319220705	20523920152	22329629382	22246977807
RMSE	1065870606	917321409	813259007	820978567	810300852
MAPE	0.039842	0.038842	0.030115	0.025444	0.024715

According to Table **3**, DC-ANN produced the most accurate forecasts for Turkey electricity consumption data in terms of both RMSE and MAPE since this method has the best performance measure values.

CONCLUSION

In this paper, a new neural network model is proposed. For two real world data sets, DC-ANN method can produce the most accurate forecast results according to RMSE and MAPE criteria. As a result of this applications, it can be said that deterministic components can be useful as inputs of neural networks. Moreover, the proposed network is a combined network of regression models and multiplicative neuron model and the combination weights are also weights of DC-ANN. This combination style can be applied for other time series forecasting models in the literature, this can be a good research idea in the future. In the future studies, DC-ANN can be modified as recurrent version.

CONFLICT OF INTEREST

The authors (editor) declare no conflict of interest, financial or otherwise.

ACKNOWLEDGEMENTS

Declared none.

REFERENCES

[1]   J.P. Teixeira, and P.O. Fernandes, "Tourism Time Series Forecast- Different ANN Architectures with Time Index Input", *Procedia Technology,* vol. 5, pp. 445-454, 2012.
[http://dx.doi.org/10.1016/j.protcy.2012.09.049]

[2]   S. Makridakis, S.C. Wheelwright, and R.J. Hyndman, *Forecasting Methods and Applications.* 3rd ed. Wiley & Sons: New York, 1998.

[3]   G.E. Box, and G.M. Jenkins, *Time Series Analysis: Forecasting and Control.* Holdan-Day: San Francisco, 1976.

[4]   W.W. Wei, *Time Series Analysis: Univariate and Multivariate Methods.* 2nd ed. Pearson Addison Wesley: USA, 2006.

[5]   D.E. Rumelhart, G.E. Hinon, and R.J. Williams, "Learning representations by backpropagation error", *Nature,* vol. 32, pp. 533-536, 1986.
[http://dx.doi.org/10.1038/323533a0]

[6]   J.L. Elman, "Finding structure in time", *Cogn. Sci.,* vol. 14, pp. 179-211, 1990.
[http://dx.doi.org/10.1207/s15516709cog1402_1]

[7]   R.N. Yadav, P.K. Kalra, and J. John, "Time series prediction with single multiplicative neuron model", *Appl. Soft Comput.,* vol. 7, pp. 1157-1163, 2007.
[http://dx.doi.org/10.1016/j.asoc.2006.01.003]

[8]   L. Zhao, and Y. Yang, "PSO-based single multiplicative neuron model for time series prediction", *Expert Syst. Appl.,* vol. 36, pp. 2805-2812, 2009.
[http://dx.doi.org/10.1016/j.eswa.2008.01.061]

[9]   B. Samanta, "Prediction of chaotic time series using computational intelligence", *Expert Syst. Appl.,* vol. 38, no. 9, pp. 11406-11411, 2011.
[http://dx.doi.org/10.1016/j.eswa.2011.03.013]

[10]   U. Yolcu, C.H. Aladag, and E. Egrioglu, "A New Linear & Nonlinear Artificial Neural Network Model for Time Series Forecasting", *Decision Support System Journals,* vol. 54, pp. 1340-1347, 2013.
[http://dx.doi.org/10.1016/j.dss.2012.12.006]

[11]   C.H. Aladag, U. Yolcu, and E. Egrioglu, "A new multiplicative seasonal neural network model based on particle swarm optimization", *Neural Process. Lett.,* vol. 37, no. 3, pp. 251-262, 2013.
[http://dx.doi.org/10.1007/s11063-012-9244-y]

[12]   E. Egrioglu, U. Yolcu, C.H. Aladag, and E. Bas, "Recurrent Multiplicative Neuron Model Artificial Neural Network for Non-Linear Time Series Forecasting", *Neural Process. Lett.,* vol. 41, no. 2, pp. 249-258, 2015.
[http://dx.doi.org/10.1007/s11063-014-9342-0]

[13]   F. Van Den Bergh, and A.P. Engelbrech, "A new locally convergent particle swarm optimiser", *Systems, Man and Cybernetics, 2002 IEEE International Conference on,* vol. 3, 2002.
[http://dx.doi.org/10.1109/ICSMC.2002.1176018]

[14]   G.J. Janacek, *Practical time series.* Oxford University Press: New York, 2001.

CHAPTER 5

A Fuzzy Time Series Approach Based on Genetic Algorithm with Single Analysis Process

Ozge Cagcag Yolcu[*]

Department of Industrial Engineering, Faculty of Engineering, Giresun University, Giresun, Turkey

Abstract: In the literature, two basic approaches are mentioned for time series forecasting. These are probabilistic and non-probabilistic approaches. This study is focused on fuzzy time series method one of the non-probabilistic approaches. Fuzzy time series analysis methods are the effective methods which are more favourable than traditional methods. The basic stages as fuzzification, identification of fuzzy relations and defuzzification which constitute the fuzzy time series analyses has been affectively used to get a better prediction performance. All of these three stages that are considered separately in analysis process lead to different errors. This situation, therefore, may cause a rise in model error. In order to eliminate this problem in this study all steps can be evaluated in one process synchronously. In the proposed approach, the method similar to fuzzy C-means, multiplicative artificial neural networks and genetic algorithm are used simultaneously in fuzzification, identification of fuzzy relation and determination of all parameters, respectively. And also different fuzzy time series are analysed. All obtained results are discussed to be able to consider the proposed method in terms of forecasting performance.

Keywords: Fuzzy Time Series, Forecasting, Genetic algorithm, Single multiplicative neuron model.

INTRODUCTION

Forecasting is an indispensable element in the decision making process which can be considered as making inference about events or conditions in the future by using existing information. The aim of time series analysis is to obtain accurate predictions for the future. If a model which is suitable for a time series can be specified, the future values of this time series can be predicted by using this model. Actually, we can say that everything in the world can be taken into account as a time series. In this respect, time series occur in a variety of fields such as business and economics, engineering, agriculture, geophysics, medicine,

[*] **Corresponding author Ozge Cagcag Yolcu:** Giresun University, Faculty of Engineering, Department of Industrial Engineering, Giresun, 28200, Turkey; E-mail: ozgecagcag@yahoo.com.

Cagdas Hakan Aladag (Ed.)

meteorology, quality control and social science. So, numerous approaches have been proposed in scientific literature during the last few decades. But in some situations requiring some strict assumptions in classical time series may lead to researchers to use FTS methods. These basic assumptions are normality of distributions of residuals, homogeneity of variance and zero mean of residuals, independence of residuals, absence of outliers, stationarity and invertibility and linearity model. In addition, not requiring any assumption in forecasting time series makes advanced time series which are called non-stochastic models, forecasting methods applicable for many fields such as ANN and Fuzzy Approaches.

Nowadays, as a family of non-stochastic approaches, FTS methods become prominent in time series forecasting because of their some advantages. In the literature, Song and Chissom [1] was proposed first fuzzy time series (FTS) study based on Zadeh's fuzzy set theory [2].

Egrioglu *et al.* [31] said that FTS approaches do not require to be satisfied a variety of assumptions in opposition to traditional time series analysis methods. Having the ability of working with very small data sets is the most essential advantage of the FTS approaches. Egrioglu *et al.* [30] also stated that most of data sets encountered in time series analysis problems, such as *air temperature, air quality, stock index and exchange rate because of the vagueness that they contain,* should be examined as FTS. *On the occasion of all these reasons, we can say that the interest in FTS approach is progressively increasing.*

As all we know time-variant and time-invariant concepts should be able to examine for analysing of FTS methods. While it is assumed that relation of time series changes over time in the time-variant FTS, relation of time series does not change over time in the time invariant FTS. Although there are studies related to both time-variant and time invariant FTS in the literature, most of them are about time-invariant FTS. And also, in this study time-invariant FTS is viewed. Fundamentally, the methods of FTS are based on three stages as fuzzification, identification of fuzzy relations and defuzzification. Each of these stages has an effective role on forecasting performance of the methods. On these three steps, many studies using different approaches have been carried out by various researchers.

In the fuzzification step, partition of universe of discourse has common usage area. For this reason, in the literature, there are a great number of studies about determining of interval length. In most of these studies, interval lengths are determined constantly but Song and Chissom [1, 3, 4] and Chen [5, 6] arbitrarily fixed equal interval lengths whereas Huarng [7] made benefit of two different approaches as average and distribution based and Egrioglu and co-workers [8, 9] used optimization based methods. To determine the dynamic intervals, while Huarng and Yu [10] and Yolcu and co-workers [11] used the approaches based on ratio, Cheng and Chung [12], Lee and co-workers [13, 14], Kuo and co-workers [15, 16], Davari and co-workers [17], Park and co-workers [18], Hsu and co-workers [19], and Yolcu and co-workers [20] utilized the artificial intelligent optimization techniques such as particle swarm optimization (PSO), genetic algorithm (GA) and artificial bee colony algorithm. In addition, Cheng and co-workers [21], Li and co-workers [22], Aladag and co-workers [23], Alpaslan and co-workers [24], Egrioglu [25] and Egrioglu and co-workers [26] benefited from Fuzzy C-Means (FCM) clustering technique to avoid subjective judgements.

In the first years, complex matrix operations were used by Song and Chissom [1, 3, 4] to be able to determine the fuzzy relations. Chen [5] proposed a new model that involves easier operations based on fuzzy group relation tables. In addition, in recent years, more systematic approaches have been adopted by avoiding the complex methods. Huarng and Yu [27], Aladag and co-workers [28, 29], Egrioglu and co-workers [30 - 32] proposed approaches using multilayer perceptron feed-forward artificial neural networks (MLP-FF-ANN) to identify fuzzy relations. One of the basic problems of MLP-FF-ANN is determining of the number of neuron in the hidden layer. To avoiding this problem, Aladag [33] used single multiplicative neuron model artificial neural networks (SMNM-ANN) introduced by Yadav and co-workers [34]. In all of these approaches, while the fuzzy relations which indicate the internal relation of FTS is specifying, the fuzzy set which has the highest membership value is considered and the others are disregarded. In this case information loss can be come out and the negative influence on methods will be inevitable.To be able to cope with the problem, Yu and Huarng [35, 36] used forecasting models in which membership values are intuitively determined. Alpaslan and Cagcag [37], Alpaslan and co-workers [24], and Yolcu and co-workers [38] used FCM technique rather than identification of the membership values intuitively.

In the defuzzification stage, centroid method has been used in almost all studies. Moreover Jilani and Burney [39, 40] and Jilani and co-workers [41] proposed different approaches in this stage.

Each of these three stages that are considered separately in analysis process leads to different errors. This situation, therefore, may cause rising of model error. A way of avoiding from this problem may be the evaluating these three stages, synchronously and simultaneously. In this study, from this viewpoint, we proposed an approach which has a single analysis process.

In this proposed method, FCM was utilized for fuzzification, SMNM-ANN was used for identification of fuzzy relations and it is unnecessary to use defuzzification operation which is applied by using real values of date as target of SMNM-ANN. In the analysis process all of the parameters of FCM and SMNM-ANN were determined in the single optimization process by using GA.

In the rest of the paper, related definitions and notions of FTS are represented. And then, GA and SMNM-ANN are briefly given. After the explanation of proposed method, the experimental results are presented. The experimental results are summarized and discussed in the last section.

FUZZY TIME SERIES

Related definitions of FTS were given by Song and Chissom [1, 3, 4] as follows:

Definition 1. Let $Y(t)$ (t = 0, 1, 2, …), a subset of real numbers, be the universe of discourse on which fuzzy sets $f_j(t)$ are defined. If $F(t)$ is a collection of $f_1(t)$, $f_2(t)$, … then $F(t)$ is called a fuzzy time series defined on $Y(t)$.

Definition 2. Suppose $F(t)$ is caused by $F(t-1)$ only, *i.e*, $F(t-1) \rightarrow F(t)$. Then this relation can be expressed as $F(t) = F(t-1)°R(t, t-1)$ where $R(t, t-1)$ is the fuzzy relationship between $F(t-1)$ and $F(t)$, and $F(t) = F(t-1)°R(t, t-1)$ is called the first order model of $F(t)$. " $°$ " represents max-min composition of fuzzy sets.

Definition 3. Suppose $R(t, t-1)$ is a first order model of $F(t)$. If for any t, $R(t, t-1)$ is independent of t, *i.e*, for any t, $R(t, t-1) = R(t-1, t-2)$, then $F(t)$ is called a time invariant fuzzy time series otherwise it is called a time variant fuzzy time series.

Song and Chissom [3] firstly suggested a first order forecasting model and its algorithm to forecast time invariant $F(t)$. In this model, the fuzzy relationship matrix $R(t, t-1)$ is obtained by many matrix operations and fuzzy forecasts are procured via max-min composition as follows:

$$F(t) = F(t-1)°R \tag{1}$$

The number of fuzzy sets composed of partition of universe of discourse specifies the dimension of R matrix. The more we use more fuzzy sets, the more we need more difficult matrix operations for obtain R matrix."

RELATED METHODS

Genetic Algorithm (GA)

Genetic algorithms imitate the process that living creatures are exposed to and based on principles that strong generations protect their life and poor generations disappear. And also, GA based on that individuals who are born from their parents keep going their life by adapting to living conditions.

Holland [42] was the prisoner of studying genetic algorithms. Population size, mutation rate, evaluation function, crossover rate and number of generation are the well-known genetic algorithm parameters. The most basic component of genetic algorithm is gene. The gens generate the chromosomes. And the population consists of the chromosomes. Each of gens represents the value of decision variable and each of chromosomes represents a point of solution space in an optimization problem. A simple population structure can be given as Fig. (**1**).

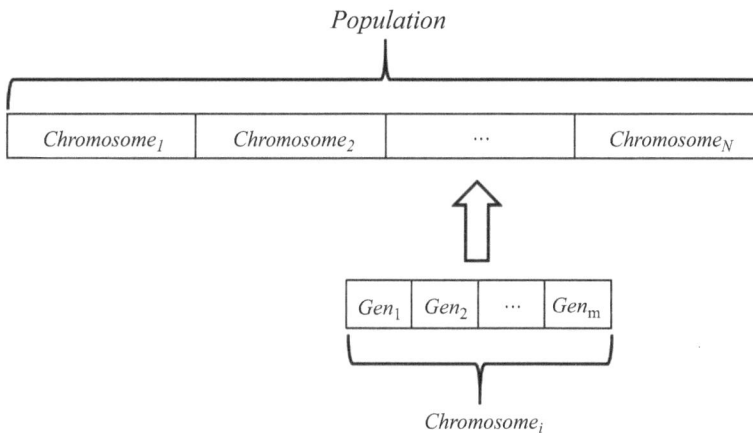

Fig. (1). A simple population structure.

In Fig. (**1**), size of population and gens number for each chromosomes are N and m, respectively. In the implementations, while the gens number is equivalent to decision variables number, the population size is determined by researcher according to the features of optimization problem. The starting population is generated randomly. After this operation, as the solution space is searching, some genetic operators are utilized to provide the variety of population such as crossover, mutation and natural selection. Aladag *et al.* [49] summarized these operators as follows:

Crossover: In this transaction called as the crossover operation, two chromosomes of the population are randomly selected and a crossover point is also randomly picked in the selected chromosomes to swap genes after the crossover point. The crossover operation performs depend on the predetermined crossover rate. If the predetermined crossover rate is smaller than or equal to the random number generated from the uniform distribution, the crossover operation is executed. From Fig. (**2**), for crossover operation, an example can be seen.

Fig. (2). Crossover operation.

Mutation: In this transaction, first of all, one chromosome has to be randomly picked. And, a random number is generated from the uniform distribution with the parameters 0 and 1. Then, the mutation operation is performed with a randomly selected gene from the chromosomes if the mutation rate is bigger than the generated random number. For the problem with different characteristic, different mutation operation can be executed. (Fig. **3**) represents an example of mutation operation.

Fig. (3). Mutation operation.

Natural Selection: Each chromosome of any generation is evaluated in terms of pre-determined evaluation function and while the best chromosomes are transferred into the next generation, some chromosomes of the worst ones are discarded. Finally, the new chromosomes are randomly generated to replace discarded ones.

Single Multiplicative Neuron Model

In neurons of feed-forward neural networks, the input signal is calculated based on addition function. Yadav and co-workers [34] proposed a new ANN model which is called single multiplicative neuron model. In their study, multiplication function is used to get the neuron's input signal. Yadav and co-workers [34] showed that single multiplicative neuron model gives better forecasting performance for forecasting of time series. Kennedy and Eberhart [44] put forward the study of PSO in the literature, and also Zhao and Yang [43] recommended the usage of PSO instead of back propagation learning algorithm proposed by Yadav and co-workers [34] in the training of single multiplicative neuron model. (Fig. **4**).represents an example of single multiplicative neuron model structure for 5 inputs.

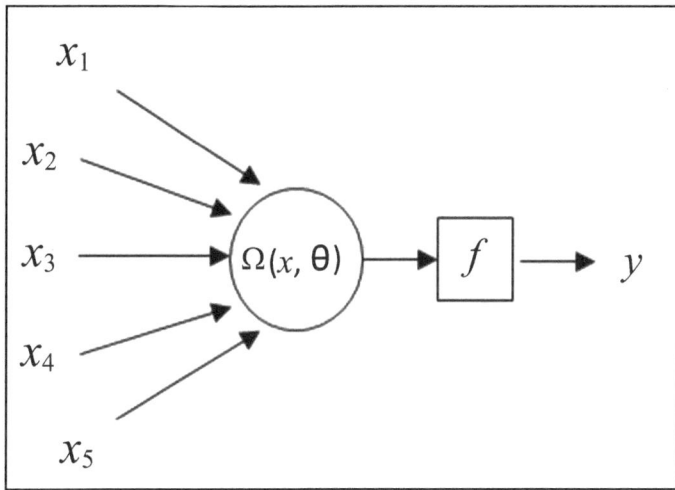

Fig. (4). An Example of Single Multiplicative Neuron Model Structure.

This model has a single neuron and the input signal of the neuron is estimated by the multiplication function. Function $\Omega(x,\Theta)$ is the product of the multiplication of weighted inputs. An example of the multiplicative neural model with five inputs given in Fig. **4** $(x_i, i = 1,2, \dots ,5)$ has 10 weights. Five of these are the weights corresponding to the inputs $(w_i, i = 1,2, \dots, 5)$ and five to the sides of the weights $(b_i, i = 1,2, \dots, 5)$. Suppose that activation function is taken as logistic function given below.

$$f(x) = \frac{1}{1+e^{-x}} \tag{2}$$

so, neuron's net value can be taken as follows.

$$net = \Omega(x,\theta) = \prod_{i=1}^{5}(w_i x_i + b_i) \tag{3}$$

After getting net value, output of weight is obtained as $y = f(net)$ by way of activation function. Sum of squares is used as a fitness function, which is calculated throughout the training of multiplicative neuron model with GA. It can be given as follow,

$$SSE = \sum_{i=1}^{n}(d_i - y_i)^2 \tag{4}$$

where d_i and y_i show the target value and the network output corresponding to i^{th} learning sample.

PROPOSED METHOD

Each of these three stages that constitutes the process of FTS analysis is taken in consideration which leads to different errors. This situation, therefore, may cause rise in model error. In this study, from this point of view, a new approach which has a single analysis process has been proposed. In the proposed method these three stages are considered synchronously and simultaneously.

In the proposed method, to fuzzification operation and to identification of fuzzy relations, the equations based on fuzzy clustering technique (FCM) and SMNM-ANN have been utilized, respectively. Moreover, it is unnecessary to use defuzzification operation which is applied by using real values of date as target of SMNM-ANN. Therefore, in the proposed method, all of the parameters of process have been determined in the single optimization process by using genetic algorithm (GA).

The membership values that compose the input of SMNM-ANN are obtained based on the center of fuzzy cluster that determined by GA in the optimization process.

$$\mu_{it} = \frac{1}{\sum_{k=1}^{c}\left(\frac{d(X(t),v_i)}{d(X(t),v_k)}\right)^{2/(\beta-1)}} , i = 1,2,\cdots,c ; \ t = 1,2,\cdots,T \tag{5}$$

Where $\beta \epsilon R$ with $(\beta \geq 1)$ is a fuzziness index and it governs the greatness of memberships. $d(X(t), v_i)$, Euclidian distance, represents a similarity measure the data and cluster centre. Satapathy *et al.* [50] define fuzzy clustering as follows:

The fuzzy matrix μ which has T (# data objects) rows and c (# clusters) columns describes the fuzzy clustering. The degree of association or membership function of the ith object with the tth cluster is represented by the element in the ith row and t^{th} column in the fuzzy matrix (μ_{it}). The characters of μ are as follows:

$$\mu_{it} \in [0,1] \quad \forall\, i = 1,2,\cdots,c \;;\; \forall\, t = 1,2,\cdots,T \tag{6}$$

$$\sum_{i=1}^{c} \mu_{it} = 1 \quad \forall\, t = 1,2,\cdots,T \tag{7}$$

$$0 < \sum_{t=1}^{T} \mu_{it} < T \quad \forall\, i = 1,2,\cdots,c \tag{8}$$

After briefly defining fuzzy clustering with the features, structure of SMNM-ANN having these features is shown in Fig. **(5)**.

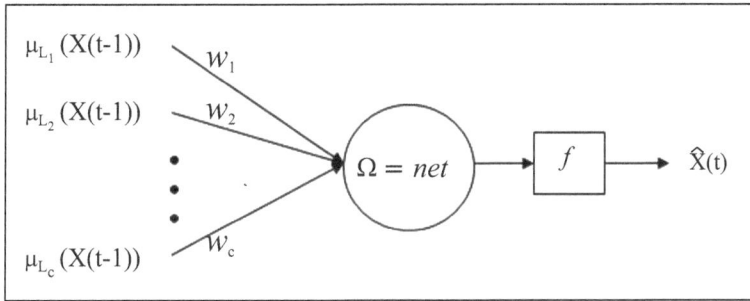

Fig. (5). The structure of SMNM-ANN.

In Fig. **(5)**, $\mu_{L_c}(X(t-1))$ is the membership value representing the degree of observation pertaining to c^{th} fuzzy sets at $(t-1)$ and makes up the network inputs. Ω function comprises multiplication of the weighted inputs and is obtained by eq. (9). In addition, f denotes the activation function whereas $\hat{X}(t)$ represents inputs of the model. Output of the model is calculated as in eq. (10).

$$\Omega(\mu,w,b) = net = \prod_{i=1}^{c}\left[w_i \times \mu_{L_i}(X(t-1)) + b_i\right] \tag{9}$$

$$\hat{X}(t) = f(net) = \frac{1}{1+exp(-net)} \tag{10}$$

In the case that the fuzzy sets number defined for the defuzzification is c, there are $3 \times c$ variables to be optimized by GA. The gens related with these variables for one chromosome can be shown as in Fig. **(6)**.

The Centers of Fuzzy Clusters		The Weights of SMNM-ANN		The Biases of SMNM-ANN	
v_1	$\cdots$	v_c	w_1	$\cdots$	w_c	b_1	$\cdots$	b_c

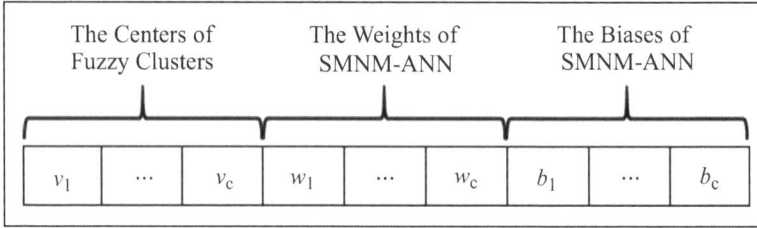

Fig. (6). The structure of a chromosome.

where, v_i, $i = 1, 2, \ldots, c$ are centers of fuzzy clusters, w_i, $i = 1, 2, \ldots, c$ and b_i, $i = 1, 2, \ldots, c$ weights and biases of SMNM-ANN, respectively.

Algorithm The proposed method.

Step 1. Define the c called number of fuzzy sets.

Sets number should be $2 < c < T$ range where T is the number of observation.

Step 2. Define the parameters of GA algorithm. These parameters can be given as follows;

Population size: Ps

Crossover rate: Cr

Mutation rate: Mr

Maximum generation: Mg

The next generation is constituted transformed best chromosomes number: NsN

Step 3. Initial generation are randomly generated.

The initial gens for centers of fuzzy clusters are randomly obtained from uniform distribution $(min(X(t)), max(X(t)))$, weights and biases of SMNM-ANN are also obtained from uniform distribution $(0,1)$ randomly.

Step 4. Obtain evaluation function values for each chromosome.

To be able to evaluate the results root mean square error (RMSE) is utilized which is given in below.

$$RMSE = \sqrt{\frac{1}{T}\sum_{t=1}^{T}\left(X(t) - \hat{X}(t)\right)^2} \qquad (11)$$

where T shows the learning sample numbers for SMNM-ANN.

Step 5. The crossover operation is carried out.

In this step firstly an initial number is randomly obtained from the uniform distribution. After that, if this number is bigger than the rate of crossover, transaction of crossover is carried out. An operation of crossover can be given as Fig. (**7**).

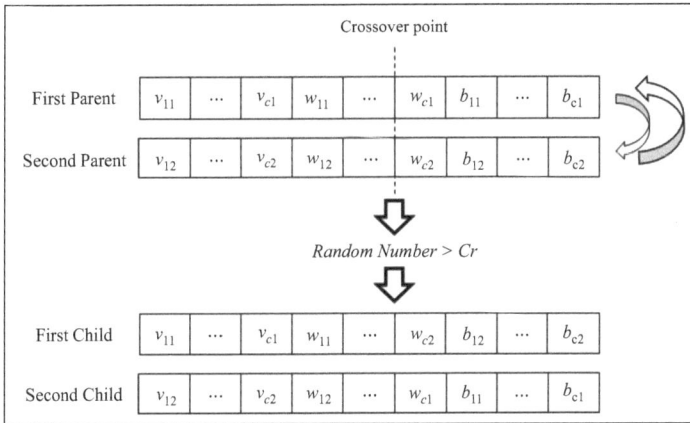

Fig. (7). A crossover operation.

Step 6. The mutation operation is carried out.

If the mutation rate is bigger than a random number generated from the uniform distribution with the parameters 0 and 1, the mutation operation will be carried out with a randomly selected gene from the chromosomes. A mutation operation can be given as Fig. (**8**).

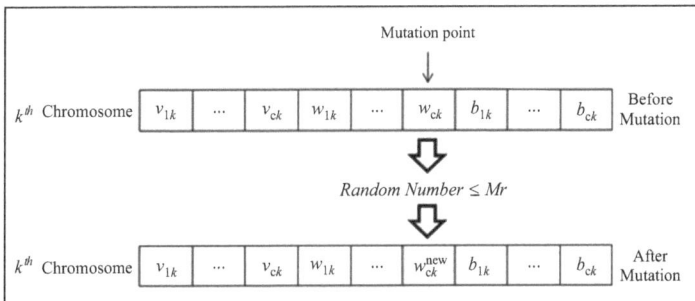

Fig. (8). A mutation operation.

Step 7. Natural selection operation is performed.

The best chromosomes are transferred into the next generation according to a rate which is determined by researcher.

Step 4-7 are repeated the number of maximum generation (Mg). Finally, the gens of chromosome which has the best value of evaluation function are taken as the optimal solution.

APPLICATIONS

In order to evaluate the performance of the proposed method, Taiwan stock index (TAIEX) data sets in 2003 and 2004 were analysed in the implementation. In the analysis process, last two and three months data of TAIEX were taken as test data (test data 1 and test data 2).

In the implementation, the parameters are taken as below:

• Fuzzy sets number(c) is changed between 5 and 25

• $Ps = 30$

• $Cr = 0.8$

• $Mr = 0.01$

• $Mg = 300$

• $NsN = 10$

The graph TAIEX data of 2003 and 2004 years can be seen in Figs. (**9-10**).

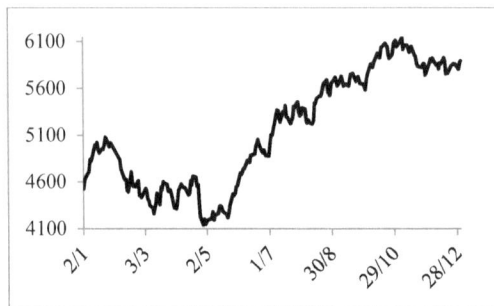

Fig. (9). TAIEX-2003 data.

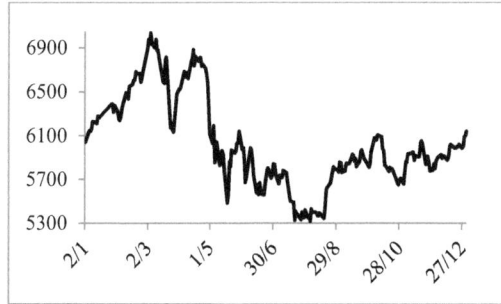

Fig. (10). TAIEX-2004 data.

For TAIEX-2003 and 2004 data, the optimal results are obtained for seven and thirteen fuzzy sets for test data 1 and 2, respectively. The best forecasting results of the proposed method and some well- known models are presented in Tables **1-2**.

Table 1. Performance evaluation of methods for TAIEX-2003 and TAIEX-2004 test data 1.

Methods	RMSE	
	2003	2004
Yu and Huarng [35]	58.00	67.00
Chen and Chang [45]	53.51	60.48
Chen and Chen [46]	58.06	57.73
Chen and co-workers [47]	52.49	52.27
The Proposed Method	49.96	51.13

Table 2. Performance evaluation of methods for TAIEX-2003 and TAIEX-2004 test data 2.

Methods	RMSE	
	2003	2004
Chen [5]	74.46	84.28
Huarng and Yu [27]	56.00	72.35
Huarng and co-workers [48]	70.76	55.91
Yu and Huarng [36]	58.78	53.83
The Proposed Method	54.47	52.47

When we view the Tables **1-2**, it is clearly seen that forecasting performance of the proposed method is better than other methods in the literature from the point of RMSE evaluation criterion for both of data. The line graphs of forecasts obtained from the proposed method and the real observations are represented in Figs. (**11-14**).

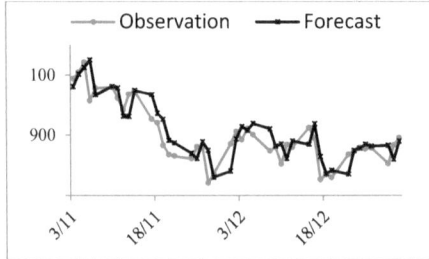

Fig. (11). The out-of-sample forecasts and observations for TAIEX-2003 test data 1.

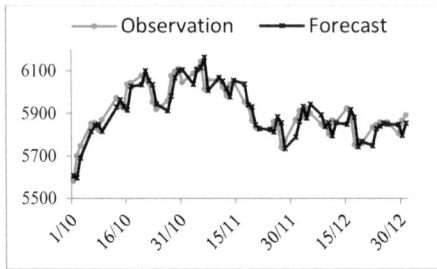

Fig. (12). The out-of-sample forecasts and observations for TAIEX-2003 test data 2.

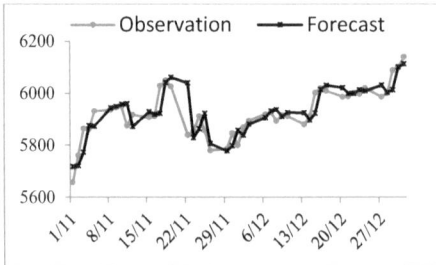

Fig. (13). The out-of-sample forecasts and observations for TAIEX-2004 test data 1.

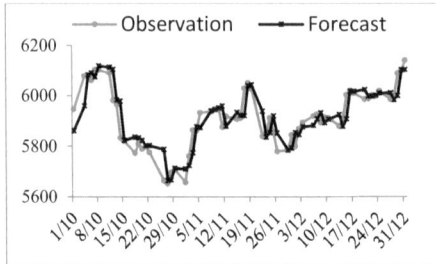

Fig. (14). The out-of-sample forecasts and observations for TAIEX-2004 test data 2.When Figs. (**11-14**) are considered, it can be seen that the forecast are conformed to the real observations all of test data.

CONCLUSION AND DISCUSSION

Nowadays, as a family of non-probabilistic approaches, FTS methods become prominent in terms of time series forecasting. While the FTS methods which have outstanding features in different aspects were presented in the literature, they still need to be improved in many aspects. The separately examining of analysis process can be seen as one of them. This study focused on to solve this problem. In this study, GA was also used as an optimization tool as well as analysis process was taken into consideration as single optimization process. That is, the parameters of fuzzification and the identification of fuzzy relations process are obtained in a single process by using GA. While the parameters of the fuzzification procedure are based on equations of FCM, the parameters of the identification of fuzzy relations procedure are composed of weights and biases of SMN-ANN. The proposed method was implemented two time series and attained results are compared with some other studies results. When the obtained findings are evaluated, and impressive forecasting performance of the method must be emphasized.

In the future studies, it can be researched the influence of using different artificial intelligent optimization techniques and artificial neural networks. In addition, the presented model in this study can be transform into a high-order model.

CONFLICT OF INTEREST

The author confirms that they have no conflict of interest to declare for this publication.

ACKNOWLEDGEMENTS

Declared none.

REFERENCES

[1]   Q. Song, and B.S. Chissom, "Fuzzy time series and its models", *Fuzzy Sets Syst.,* vol. 54, pp. 269-277, 1993.
        [http://dx.doi.org/10.1016/0165-0114(93)90372-O]

[2]   L.A. Zadeh, "Fuzzy Sets", *Inf. Control,* vol. 8, pp. 338-353, 1965.
        [http://dx.doi.org/10.1016/S0019-9958(65)90241-X]

[3]   Q. Song, and B.S. Chissom, "Forecasting enrollments with fuzzy time series- Part I", *Fuzzy Sets Syst.,* vol. 54, pp. 1-10, 1993.
        [http://dx.doi.org/10.1016/0165-0114(93)90355-L]

[4]   Q. Song, and B.S. Chissom, "Forecasting enrollments with fuzzy time series- Part II", *Fuzzy Sets Syst.,* vol. 62, pp. 1-8, 1994.
        [http://dx.doi.org/10.1016/0165-0114(94)90067-1]

[5]   S.M. Chen, "Forecasting enrolments based on fuzzy time-series", *Fuzzy Sets Syst.,* vol. 81, pp. 311-319, 1996.

[http://dx.doi.org/10.1016/0165-0114(95)00220-0]

[6]   S.M. Chen, "Forecasting enrolments based on high order fuzzy time series", *Cybern. Syst.*, vol. 33, pp. 1-16, 2002.
        [http://dx.doi.org/10.1080/019697202753306479]

[7]   K. Huarng, "Effective length of intervals to improve forecasting in fuzzy time-series", *Fuzzy Sets Syst.*, vol. 123, pp. 387-394, 2001.
        [http://dx.doi.org/10.1016/S0165-0114(00)00057-9]

[8]   E. Egrioglu, C.H. Aladag, U. Yolcu, V.R. Uslu, and M.A. Basaran, "Finding an optimal interval length in high order fuzzy time series", *Expert Syst. Appl.*, vol. 37, pp. 5052-5055, 2010.
        [http://dx.doi.org/10.1016/j.eswa.2009.12.006]

[9]   E. Egrioglu, C.H. Aladag, M.A. Basaran, V.R. Uslu, and U. Yolcu, "A New Approach Based on the Optimization of the Length of Intervals in Fuzzy Time Series", *J. Intell. Fuzzy Syst.*, vol. 22, pp. 15-19, 2011.

[10]   K. Huarng, and T.H Yu, "Ratio-based lengths of intervals to improve fuzzy time series forecasting", *IEEE Trans. Syst. Man Cybern. B Cybern.*, vol. 36, pp. 328-340, 2006.
        [http://dx.doi.org/10.1109/TSMCB.2005.857093]

[11]   U. Yolcu, E. Egrioglu, V.R. Uslu, M.A. Basaran, and C.H. Aladag, "A New Approach for Determining the Length of Intervals for Fuzzy Time Series", *Appl. Soft Comput.*, vol. 9, pp. 647-651, 2009.
        [http://dx.doi.org/10.1016/j.asoc.2008.09.002]

[12]   S.M. Chen, and N.Y. Chung, "Forecasting enrolments using high order fuzzy time series and genetic algorithms", *Int. J. Intell. Syst.*, vol. 21, pp. 485-501, 2006.
        [http://dx.doi.org/10.1002/int.20145]

[13]   L.W. Lee, L.H. Wang, and S.M. Chen, "Temperature prediction and TAIFEX forecasting based on fuzzy logical relationships and genetic algorithms", *Expert Syst. Appl.*, vol. 33, pp. 539-550, 2007.
        [http://dx.doi.org/10.1016/j.eswa.2006.05.015]

[14]   L.W. Lee, L.H. Wang, and S.M. Chen, "Temperature prediction and TAIFEX forecasting based on high-order fuzzy logical relationships and genetic simulated annealing techniques", *Expert Syst. Appl.*, vol. 34, pp. 328-336, 2008.
        [http://dx.doi.org/10.1016/j.eswa.2006.09.007]

[15]   I-H. Kuo, S-J. Horng, T-W. Kao, T-L. Lin, C-L. Lee, and Y. Pan, "An improved method for forecasting enrollments based on fuzzy time series and particle swarm optimization", *Expert Syst. Appl.*, vol. 36, pp. 6108-6117, 2009.
        [http://dx.doi.org/10.1016/j.eswa.2008.07.043]

[16]   I-H. Kuo, S-J. Horng, Y-H. Chen, R-S. Run, T-W. Kao, R-J. Chen, J-L. Lai, and T-L. Lin, "Forecasting TAIFEX based on fuzzy time series and particle swarm optimization", *Expert Syst. Appl.*, vol. 37, pp. 1494-1502, 2010.
        [http://dx.doi.org/10.1016/j.eswa.2009.06.102]

[17]   S. Davari, M.H. Zarandi, and I.B. Turksen, "An Improved fuzzy time series forecasting model based on particle swarm intervalization", *Proc. The 28th North American Fuzzy Information Processing Society Annual Conferences (NAFIPS 2009)* Cincinnati, Ohio, USA.
        [http://dx.doi.org/10.1109/NAFIPS.2009.5156420]

[18]   J-I. Park, D-J. Lee, C-K. Song, and M-G. Chun, "TAIFEX and KOSPI 200 forecasting based on two factors high order fuzzy time series and particle swarm optimization", *Expert Syst. Appl.*, vol. 37, pp. 959-967, 2010.
        [http://dx.doi.org/10.1016/j.eswa.2009.05.081]

[19]   L-Y. Hsu, S-J. Horng, T-W. Kao, Y-H. Chen, R-S. Run, R-J. Chen, J-L. Lai, and I-H. Kuo, "Temperature prediction and TAIFEX forecasting based on fuzzy relationships and MTPSO

techniques", *Expert Syst. Appl.,* vol. 37, pp. 2756-2770, 2010.
[http://dx.doi.org/10.1016/j.eswa.2009.09.015]

[20] U. Yolcu, O. Cagcag, C.H. Aladag, and E. Egrioglu, "An enhanced fuzzy time series forecasting method based on artificial bee colony", *J. Intell. Fuzzy Syst.,* vol. 26, no. 6, pp. 2627-2637, 2014.

[21] C-H. Cheng, G-W. Cheng, and J-W. Wang, "Multi-attribute fuzzy time series method based on fuzzy clustering", *Expert Syst. Appl.,* vol. 34, pp. 1235-1242, 2008.
[http://dx.doi.org/10.1016/j.eswa.2006.12.013]

[22] S-T. Li, Y-C. Cheng, and S-Y. Lin, "A FCM-based deterministic forecasting model for fuzzy time series", *Comput. Math. Appl.,* vol. 56, pp. 3052-3063, 2008.
[http://dx.doi.org/10.1016/j.camwa.2008.07.033]

[23] S. Aladag, C.H. Aladag, T. Mentes, and E. Egrioglu, "A New Seasonal Fuzzy Time Series Method Based On The Multiplicative Neuron Model And SARIMA", *Hacet. J. Math. Stat.,* vol. 41, no. 3, pp. 337-345, 2012.

[24] F. Alpaslan, O. Cagcag, C.H. Aladag, U. Yolcu, and E. Egrioglu, "A novel seasonal fuzzy time series method", *Hacet. J. Math. Stat.,* vol. 4, no. 3, pp. 375-385, 2012.

[25] E. Egrioglu, "A New Time-Invariant Fuzzy Time Series Forecasting Method Based on Genetic Algorithm", *Advances in Fuzzy Systems,* vol. 2012, 2012.
[http://dx.doi.org/10.1155/2012/785709]

[26] E. Egrioglu, C.H. Aladag, and U. Yolcu, "Fuzzy time series forecasting with a novel hybrid approach combining fuzzy c-means and neural networks", *Expert Syst. Appl.,* vol. 40, pp. 854-857, 2013.
[http://dx.doi.org/10.1016/j.eswa.2012.05.040]

[27] K. Huarng, and H-K. Yu, "The application of neural networks to forecast fuzzy time series", *Physica A,* vol. 363, pp. 481-491, 2006.
[http://dx.doi.org/10.1016/j.physa.2005.08.014]

[28] C.H. Aladag, M.A. Basaran, E. Egrioglu, U. Yolcu, and V.R. Uslu, "Forecasting in high order fuzzy time series by using neural networks to define fuzzy relations", *Expert Syst. Appl.,* vol. 36, pp. 4228-4231, 2009.
[http://dx.doi.org/10.1016/j.eswa.2008.04.001]

[29] C.H. Aladag, U. Yolcu, and E. Egrioglu, "A high order fuzzy time series forecasting model based on adaptive expectation and artificial neural networks", *Math. Comput. Simul.,* vol. 1, pp. 875-882, 2010.
[http://dx.doi.org/10.1016/j.matcom.2010.09.011]

[30] E. Egrioglu, C.H. Aladag, U. Yolcu, M.A. Basaran, and V.R. Uslu, "A new hybrid approach based on SARIMA and partial high order bivariate fuzzy time series forecasting model", *Expert Syst. Appl.,* vol. 36, pp. 7424-7434, 2009.
[http://dx.doi.org/10.1016/j.eswa.2008.09.040]

[31] E. Egrioglu, C.H. Aladag, U. Yolcu, V.R. Uslu, and M.A. Basaran, "A new approach based on artificial neural networks for high order multivariate fuzzy time series", *Expert Syst. Appl.,* vol. 36, pp. 10589-10594, 2009.
[http://dx.doi.org/10.1016/j.eswa.2009.02.057]

[32] E. Egrioglu, V.R. Uslu, U. Yolcu, M.A. Basaran, and C.H. Aladag, "A new approach based on artificial neural networks for high order bivariate fuzzy time series", In: *Applications of Soft Computing, AISC 58,* J. Mehnen, Ed., Springer-Verlag Berlin : Heidelberg, 2009, pp. 265-273.
[http://dx.doi.org/10.1007/978-3-540-89619-7_26]

[33] C.H. Aladag, "Using multiplicative neuron model to establish fuzzy logic relationships", *Expert Syst. Appl.,* vol. 40, no. 3, pp. 850-853, 2012.
[http://dx.doi.org/10.1016/j.eswa.2012.05.039]

[34] R.N. Yadav, P.K. Kalra, and J. John, "Time series prediction with single multiplicative neuron model", *Appl. Soft Comput.,* vol. 7, pp. 1157-1163, 2007.

[http://dx.doi.org/10.1016/j.asoc.2006.01.003]

[35]   H-K. Yu, and K. Huarng, "A bivariate fuzzy time series model to forecast TAIEX", *Expert Syst. Appl.,* vol. 34, pp. 2945-2952, 2008.
[http://dx.doi.org/10.1016/j.eswa.2007.05.016]

[36]   H-K. Yu, and K. Huarng, "A neural network- based fuzzy time series model to improve forecasting", *Expert Syst. Appl.,* vol. 37, pp. 3366-3372, 2010.
[http://dx.doi.org/10.1016/j.eswa.2009.10.013]

[37]   F. Alpaslan, and O. Cagcag, "A Seasonal Fuzzy Time Series Forecasting Method Based on Gustafson-Kessel Fuzzy Clustering", *Journal of Social and Economic Statistics,* vol. 1, pp. 1-13, 2012.

[38]   U. Yolcu, C.H. Aladag, E. Egrioglu, and V.R. Uslu, "Time series forecasting with a novel fuzzy time series approach: an example for Istanbul stock market", *J. Stat. Comput. Simul.,* vol. 83, no. 4, pp. 597-610, 2013.
[http://dx.doi.org/10.1080/00949655.2011.630000]

[39]   T.A. Jilani, and S.M. Burney, "M-factor high order fuzzy time series forecasting for road accident data: Analysis and design of intelligent systems using soft computing techniques", *Advances in Soft Computing,* vol. 41, pp. 246-254, 2007.
[http://dx.doi.org/10.1007/978-3-540-72432-2_25]

[40]   T.A. Jilani, and S.M. Burney, "Multivariate stochastic fuzzy forecasting models", *Expert Syst. Appl.,* vol. 35, no. 3, pp. 691-700, 2008.
[http://dx.doi.org/10.1016/j.eswa.2007.07.014]

[41]   T.A. Jilani, S.M. Burney, and C. Ardil, "Multivariate high order fuzzy time series forecasting for car road accidents", *International Journal of Computational Intelligence.,* vol. 4, no. 1, pp. 15-20, 2007.

[42]   J.H. Holland, *Adaptation in natural and artificial systems.* MIT Press: Cambridge, MA, 1975.

[43]   L. Zhao, and Y. Yang, "PSO-based single multiplicative neuron model for time series prediction", *Expert Syst. Appl.,* vol. 36, pp. 2805-2812, 2009.
[http://dx.doi.org/10.1016/j.eswa.2008.01.061]

[44]   J. Kennedy, and R.C. Eberhart, "Particle Swarm Optimization", *Proceedings of IEEE International Conference on Neural Networks,* vol. 4, pp. 1942-1948, 1995.
[http://dx.doi.org/10.1109/ICNN.1995.488968]

[45]   S.M. Chen, and Y.C. Chang, "Multi-variable fuzzy forecasting based on fuzzy clustering and fuzzy rule interpolation techniques", *Inf. Sci.,* vol. 180, no. 24, pp. 4772-4783, 2010.
[http://dx.doi.org/10.1016/j.ins.2010.08.026]

[46]   S.M. Chen, and C.D. Chen, "TAIEX forecasting based on fuzzy time series and fuzzy variation groups", *IEEE Trans. Fuzzy Syst.,* vol. 19, no. 1, pp. 1-12, 2011.
[http://dx.doi.org/10.1109/TFUZZ.2010.2073712]

[47]   S.M. Chen, H.P. Chu, and T.W. Sheu, "TAIEX Forecasting Using Fuzzy Time Series and Automatically Generated Weights of Multiple Factors", *Transactions on Systems, Man, and Cybernetics--Part A: Systems and Humans,* vol. 42, no. 6, pp. 1485-1495, 2012.
[http://dx.doi.org/10.1109/TSMCA.2012.2190399]

[48]   K. Huarng, H-K. Yu, and Y.W. Hsu, "A multivariate heuristic model for fuzzy time-series forecasting", *IEEE Trans. Syst. Man Cybern. B Cybern.,* vol. 37, no. 4, pp. 836-846, 2007.
[http://dx.doi.org/10.1109/TSMCB.2006.890303]

[49]   C.H. Aladag, U. Yolcu, E. Egrioglu, and E. Bas, "Fuzzy lagged variable selection in fuzzy time series with genetic algorithms", *Appl. Soft Comput.,* vol. 22, pp. 465-473, 2014.
[http://dx.doi.org/10.1016/j.asoc.2014.03.028]

[50]   S.C. Satapathy, S.K. Patnaik, C.D. Dash, and S. Sahoo, "Data Clustering Using Modified Fuzzy-PSO (MFPSO)", In: *Lecture Notes in Computer Science,* vol. 7080. Springer, 2011, pp. 136-146.

CHAPTER 6

Forecasting Stock Exchanges with Fuzzy Time Series Approach Based on Markov Chain Transition Matrix

Cagdas Hakan Aladag[1,*] and **Hilal Guney**[2]

[1] *Department of Mechanical and Industrial Engineering, University of Toronto, Toronto, Canada*

[2] *Department of Statistics, Gazi University, Ankara, Turkey*

Abstract: Stock exchanges forecasting is a popular research topic that is attracting more and more attention from researchers and practitioners. Since it is a well-known fact that stock exchanges time series include uncertainty, using conventional time series methods can lead to misleading results. Therefore, a proper approach should be employed for analysis according to the nature of the data. In the literature, fuzzy time series models have been successfully used to forecast real world time series which include vagueness. In order to handle uncertainty in stock exchanges, fuzzy time series forecasting approach proposed by Tsaur [17] is utilized in this study. Fuzzy time series forecasting model suggested by Tsaur [17] uses Markov chain transition matrix for fuzzy inference. In the implementation, the fuzzy time series forecasting model is applied to index 100 in stocks and bonds exchange market of İstanbul in order to show the performance of the model. It is seen that the forecasting model gives accurate forecasts for the data.

Keywords: Fuzzy inference, Fuzzy relations, Fuzzy time series, Markov chain transition matrix, Number of fuzzy set, Stock exchanges.

INTRODUCTION

Fuzzy set theory firstly proposed by Zadeh [14] makes it possible to work under uncertainty. Fuzzy logic is based on the extension of the rules of conventional logic. Therefore, it is possible to obtain more satisfactory solutions for real world problems. One of these real world problems is time series forecasting. Fuzzy logic based methods to analyze time series include vagueness are called fuzzy time series. It has been shown in the literature that fuzzy time series produce very accurate results for real world time series [3]. Therefore, fuzzy time series approach is getting more and more attention in recent years [2].

---

* **Corresponding author Cagdas Hakan Aladag**: Department of Mechanical and Industrial Engineering, University of Toronto, Toronto, Canada; E-mail addresses: chaladag@gmail.com

Fuzzy time series approach was firstly put forward by Song and Chissom [15, 16]. A first order fuzzy time series forecasting model was introduced in [15, 16]. In a first order fuzzy time series model, an observation is only effected by previous observation. Such a model is constructed based on only this relation between sequential observations. Fuzzy time series analysis is sort of fuzzy system. In fuzzy time series, outputs and targets of the fuzzy system are forecasts and corresponding observations, respectively. Fuzzy systems are generally composed of three basic stages such as fuzzification, fuzzy inference and defuzzification [6].

Fuzzification, fuzzy inference and defuzzification phases directly affects the performance of fuzzy time series. Therefore, there have been various studies in which it is tried to make contributions to these phases in order to get better forecasting results [4]. Chen [18] suggested a more simple fuzzy time series model by dealing with the computational complexity problem of the initial model proposed by Song and Chissom [15, 16]. Chen [18] suggested to use fuzzy logic group relation tables in fuzzy inference stage to reduce the computational complexity. Although calculations in first order fuzzy time series approach proposed by Chen [18] is easy, there are some drawbacks in this approach. Later on, various fuzzy time series forecasting models have been proposed in the literature in order to obtain better results by getting over these drawbacks of the method [8].

There have been some studies that propose first order fuzzy time series forecasting models to get accurate forecasts. Some of these studies available in the literature will be given in this section. Huarng [10] extended Chen's [18] model by adding institutional information to Chen's [18] model. Huarng and Yu [9] proposed two novel approaches which are based on the average and the distribution. Yu [12] suggested that weighting repeated fuzzy relations instead of handling them only once would produce better forecasting results. In another study, Yu [13] adjusted the lengths of intervals determined during the early stages of forecasting, when the fuzzy relationships are formulated. Cheng *et al.* [7] employed adaptive expectation model in fuzzy time series. Yolcu *et al.* [19] extended Huarng's [11] ratio based methods by using constrained optimization to determine the length of intervals.

Tsaur [17] proposed another first order fuzzy time series forecasting model in which Markov chain transition matrix is used to achieve fuzzy relationship groups. This method that is applied for first degree fuzzy time series calculated observation frequencies of repeated relationships, and subsequently considered these frequencies as repetition probabilities of the relationships, and created transition matrix from state i to state j. at the final stage he corrects the forecasts through this transition matrix according to the Markov transition process. Thus, he showed superiority to many methods available in the literature in terms of forecasting performance.

In this study, Tsaur's [17] fuzzy time series forecasting model is explained. And, it is applied to index 100 in stocks and bonds exchange market of İstanbul. The time series is composed of observations between September 9th, 2014 and August 28th, 2015. The time series does not has seasonal pattern, and is not linear. Therefore, it is not possible to use the most of conventional time series models to analyze this time series. Also, it is a well-known fact that stock exchanges includes uncertainty because of the nature of the data [5]. Hence, using fuzzy time series approach instead of conventional methods would be wiser. Furthermore, it is not needed to perform some statistical hypotheses tests or to satisfy some assumptions when fuzzy time series is utilized. As a result of the application, it is observed that fuzzy time series model based on Markov chain transition matrix produces accurate forecasting results for index 100 in stocks and bonds exchange market of İstanbul.

In the following two sections, some fundamental definitions of fuzzy time series and fuzzy time series approach proposed by Tsaur [17] are presented, respectively. In the implementation section, how Tsaur's [17] model is applied to index 100 in stocks and bonds exchange market of İstanbul is given in detail. For different numbers of fuzzy set, the performance of the model is examined. All obtained mean absolute percentage error (MAPE) and root mean square error (RMSE) values are given in the application section. We conclude in the last section with a summary and a discussion of future works.

FUZZY TIME SERIES

The definition of fuzzy time series was firstly introduce by Song and Chissom [15, 16]. Fuzzy time series approaches do not require various theoretical assumptions which are needed for conventional time series analysis methods. The most important advantage of fuzzy time series approach is its ability to work with a very small set of data and no requirement for linearity assumption [1]. The basic definitions of fuzzy time series can be summarized as follows [15, 16]:

Let U be the universe of discourse, $U = \{u_1, u_2, ...,u_n\}$. A fuzzy set A of U is defined by

$$A = \frac{f_A(u_1)}{u_1} + \frac{f_A(u_2)}{u_2} + ... + \frac{f_A(u_n)}{u_n} \tag{1}$$

where f_A is the membership function of A, $f_A:U \rightarrow[0, 1]$, and $f_A(u_i)$ indicates the grade of membership of u_i in A, where $f_A(u_i)\epsilon[0, 1]$ and $1 \leq i \leq n$.

Definition 2.1. Let $Y_t \in R^1$ ($t = 0,1,2, ...$) be a time series. If $f_i(t)$ a fuzzy set in Y_t and $F(t) = \{f_1(t), f_2(t), ...\}$, then $F(t)$ is called a fuzzy time series in Y_t.

Definition 2.2. Suppose $F(t)$ is caused by $F(t-1)$ only, *i.e.*, $F(t-1) \to F(t)$. Then this relation can be expressed as $F(t) = F(t-1) \circ R(t,t-1)$ where $R(t,t-1)$ is a fuzzy relationship, and is called the first-order model of $F(t)$.

Definition 2.3. Suppose $F(t) = A_i$ is caused by $F(t-1) = A_j$, then the fuzzy logical relationship is defined as $A_i \to A_j$.

TSAUR 'S FUZZY TIME SERIES MARKOV CHAIN MODEL

Chen [18] emphasized the complexity of fuzzy matrix processes in his study and proposed a simple algorithm. In the method, first universe of discourse is defined as to cover all data, and, universe of discourse is divided into equal intervals according to the number of linguistic variables. After the data fuzzyfied with the help of fuzzy sets defined on the universe of discourse, fuzzy logical relationships are created and classification is made according to the left sides of these relationships. Then, the fuzzy forecasts are defuzzified with average method.

In Tsaur's method [17], in addition to Chen's algorithm [18], Markov chain transition matrix is used when relationship groups are obtained. This method applied for first degree fuzzy time series considers repeated relationships depending on a probability, and has the superiority in forecasted values over the many methods in the literature by making corrections in the forecasts through the transition matrix. In the study, the well-known data of student enrollment in Alabama University is utilized. In his method, universe of discourse is defined as to cover all data, and then universe of discourse is divided into equal intervals to create fuzzy sets. Following the fuzzification, the process follows two steps of operations: fuzzy logical relationships are defined, and then Markov transition probabilities matrix is created according these relationships. The forecasts obtained by using Markov transition probabilities matrix are corrected with Markov transition process diagram. This calculation process can be summarized algorithmically as follows:

1. Define the universe of discourse.
2. Divide universe of discourse into equal intervals, and define fuzzy sets.
3. Fuzzify the historical data.
4. Determine fuzzy logical relationships and groups.
5. Establish Markov state transition matrix, and then, draw transition process diagram from fuzzy logical relationships.

6. Calculate the forecasted output by using Markov state transition matrix.
7. Adjust the tendency of the forecasted values.

These steps are explained in detail as follows:

Step 1: Definition of universe of discourse: The universe of discourse is $U = [D_{min} - D_1, D_{max} + D_2]$, where D_1 and D_2 are selected as to cover all data of universe of discourse.

Step 2: Definition of fuzzy sets: Universe of discourse is divided into intervals according to the personal experience. Depending on sub-intervals of universe of discourse U, A_i ($i = 1, 2, ..., n$) fuzzy sets can be written as in the following equalities, with membership degrees.

$$A_1 = \frac{f_{A_i}(u_1)}{u_1} + \frac{f_{A_i}(u_2)}{u_2} + \frac{f_{A_i}(u_3)}{u_3} + ... + \frac{f_{A_i}(u_n)}{u_n} \tag{2}$$

Here, u_i in the denominator of fuzzy sets A_i are sub-intervals, while numbers in nominator denote membership degrees of u_i to A_i, subject to $0 \le u_i \le 1$.

Step 3: Fuzzification of data: A data should be fuzzyfied according to the biggest membership degree. If the highest membership degree of data appears in fuzzy set A_k, the corresponding data is fuzzyfied as A_k.

Step 4: Definition of fuzzy logic relationships: In this step, fuzzy logical relationships are defined between the fuzzyfied data, and then, fuzzy relationship groups are formed. Classification process is performed based on the current status of the fuzzy logical relationships.

Step 5: Establish Markov state transition matrix and transition process: Formation of Markov transition probabilities matrix and Markov transition probabilities diagram: Transition probabilities matrix P is obtained by using from the fuzzy relationships in Step 4. Defining n state for each one of the fuzzy sets, $n \times n$ dimensional matrix is produced. State transition probabilities P_{ij}, from state A_i to state A_j in one step, are calculated with equality

$$P_{ij} = \frac{M_{ij}}{M_i} \quad i, j = 1, 2, ..., n \tag{3}$$

Here, M_{ij} and M_i denote transition time in one step from state A_i to state A_j, and data amount in state A_i respectively. Thus, Markov transition probabilities matrix

P_{ij} becomes

$$P_{ij} = \begin{bmatrix} P_{11} & P_{12} & \cdots & \cdots & P_{1n} \\ P_{21} & P_{22} & \cdots & \cdots & P_{2n} \\ \cdots & \cdots & \cdots & \cdots & \cdots \\ \cdots & \cdots & \cdots & \cdots & \cdots \\ P_{n1} & P_{n2} & \cdots & \cdots & P_{nn} \end{bmatrix} \qquad (4)$$

And then, transition process diagram is drawn using the Markov transition probability matrix.

Step 6: Calculation of forecasted values: Fuzzy forecasting is conducted regarding two cases: one-to-one and one-to-many. Following rules given by Tsaur [17] are taken into account in these calculations.

- One-to-one: If the fuzzy logical relationship group of A_i is one-to-one (*i.e.*, $A_i \rightarrow A_k$, with $P_{ik} = 1$ and $P_{ij} = 0, j \neq k$), then the forecasting of $F(t)$ is m_k, the midpoint of u_k, according to the equation $F(t) = m_k P_{ik} = m_k$.
- One-to-many: If the fuzzy logical relationship group of A_j is one-to-many (*i.e.*, $A_i \rightarrow A_1, A_2, \ldots A_n, j = 1, 2, \ldots, n$), when collected data $Y(t-1)$ at time $(t-1)$ is in the state A_j, then the forecasting of $F(t)$ is equal as

$F(t) = m_1 P_{j1} + m_2 P_{j2} + \ldots + m_{j-1} P_{j(j-1)} + Y(t-1) P_{jj} + m_{j+1} P_{j(j+1)} + \ldots + m_n P_{jn}$, where $m_1, m_2, \ldots, m_{j-1}, m_j, \ldots, m_n$ are the midpoint of $u_1, u_2, \ldots, u_{j-1}, u_j, \ldots, u_n$, and m_j is substituted for $Y(t-1)$ in order to take more information from the state A_j at time $(t-1)$.

Step 7: Adjusting forecasted values: For series with small sample size estimated Markov chain matrix is usually biased, and some adjustments for the forecasting values are suggested to revise the forecasting errors for one-to-many cases. First, in a fuzzy logical group where A_i communicates with A_i and A_j for $i \neq j, j = 1, 2, \ldots, n$. If a larger state A_j is accessible from state A_i, then the forecasting value for A_j is usually underestimated because the lower state values are used for forecasting the value of A_j. On the other hand, an overestimated value should be adjusted for the forecasting value A_j because a smaller state A_j is accessible from $A_i, i, j = 1, 2, \ldots, n$. If the data happens in the state A_i, and then jumps forward to state $A_{i+k}(k \geq 2)$ or jumps backward to state $A_{i-k}(k \geq 2)$, then it is necessary to adjust the trend of the pre-obtained forecasting value. Thus, we have smoother values of forecasting.

Corrections are made by taking into account some of the following rules. To understand the rules better, first of all necessary definitions are given.

Before giving the correction rules suggested by Tsaur [17], let us give to necessary definitions.

Definition 3.1. If $P_{ij} > 0$, then state A_j is accessible from A_i, $A_i \rightarrow A_j$.

Definition 3.2. If states A_i and A_j are accessible to each other, then A_i communicates with A_j, $A_i \leftrightarrow A_j$.

Now, the corrections are made by taking into account the following rules.

- If $A_i \leftrightarrow A_j$, starting in state A_i at time $(t-1)$ as $F(t-1) = A_i$, and makes an increasing transition into state A_j at time t, $(i < j)$, then the adjusted trend value D_t is defined as $D_{t1} = (1/2)$.
- If $A_i \leftrightarrow A_j$, starting in state A_i at time $(t-1)$ as $F(t-1) = A_i$, and makes an decreasing transition into state A_j at time t, $(i > j)$, then the adjusted D_t is defined as $D_{t1} = -(1/2)$.
- If the current state is in the state A_i at time $(t-1)$ as $F(t-1) = A_i$, and makes a jump forward transition into state A_{i+s} at time t, $(1 \leq s \leq n - i)$, adjusted D_t is defined as $D_{t2} = (1/2)s$, $(1 \leq s \leq n - i)$, where l is the length that the universal discourse U must be partitioned into as n equal intervals.
- If the process is defined to be in state A_i at time $(t-1)$ as $F(t-1) = A_i$, and makes a jump-backward transition into state A_{i-v} at time t, $(1 \leq v \leq i)$, the adjusted D_t is defined as $D_{t2} = -(1/2)v$, $(1 \leq v \leq i)$.

THE IMPLEMENTATION

Forecasting and modeling for index 100 in stocks and bonds exchange market of İstanbul (IMKB 100) is an important issue in economical literature. Therefore, the fuzzy time series forecasting model proposed by Tsaur [17] is applied to this daily data. The graph of the time series is shown in Fig. (**1**). In this graph, vertical and horizontal axes represent observed values and date, respectively.

The data set composed of observations from 09.01.2014 to 08.28.2015 were used as training set, while the observations from 31.08.2015 to 04.09.2015 were used the test data includes last five observations. First of all, the time series is fuzzified. Fuzzy inference is performed by using fuzzy observation in the training set. In other words, fuzzy relationships between the fuzzified observations are defined in fuzzy inference phase. In this inference phase, Markov chain transition matrix is employed according to the fuzzy time series approach suggested by Tsaur [17]. Then, fuzzy forecasts are calculated based on constructed Markov chain transition matrix. Finally, obtained fuzzy forecasts are defuzzified. The application of

Tsaur's [17] fuzzy time series model is given step by step.

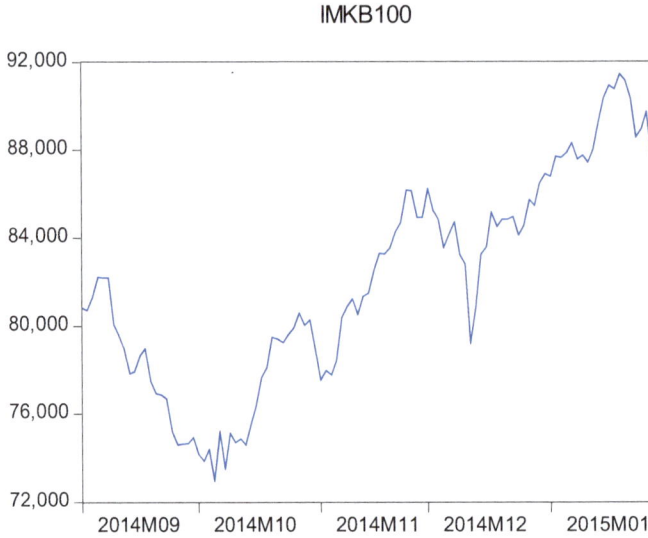

Fig. (1). Turkey IMKB 100 index daily data.

Step 1: Definition of universe of discourse:

D_{min} = 71341.95 and D_{max} = 91412.94 denote the minimum and maximum values of data set respectively. D_1 and D_2 are selected as to cover all data of universe of discourse. Thus, D_1 = 10.95 and D_2 = 10.06 are arbitrarily selected, and now the universe of discourse can be written as U = [71331, 91423].

Step 2: Definition of fuzzy sets:

Universe of discourse U = [71331, 91423] is divided into intervals according to the personal experience. Here, fuzzy sets number is changed from 5 to 15. And then, the universe of discourse is divided into equal intervals as fuzzy sets number. The subsets are given below when the number of fuzzy sets is set to 5.

$$
\begin{aligned}
u_1 &= [71331, 75349] \\
u_2 &= [75349, 79368] \\
u_2 &= [79368, 83386] \\
u_2 &= [83386, 87405] \\
u_2 &= [87405, 91423]
\end{aligned}
\tag{5}
$$

The fuzzy sets A_1, A_2, A_3, ..., A_8 depends on sub-intervals of universe of discourse U and can be written as in the following equalities with membership degrees.

$$A_1 = \frac{1}{u_1} + \frac{0.5}{u_2} + \frac{0}{u_3} + \frac{0}{u_4} + \frac{0}{u_5}$$

$$A_2 = \frac{0.5}{u_1} + \frac{1}{u_2} + \frac{0.5}{u_3} + \frac{0}{u_4} + \frac{0}{u_5}$$

$$A_3 = \frac{0}{u_1} + \frac{0.5}{u_2} + \frac{1}{u_3} + \frac{0.5}{u_4} + \frac{0}{u_5} \qquad (6)$$

$$A_4 = \frac{0}{u_1} + \frac{0}{u_2} + \frac{0.5}{u_3} + \frac{1}{u_4} + \frac{0.5}{u_5}$$

$$A_5 = \frac{0}{u_1} + \frac{0}{u_2} + \frac{0}{u_3} + \frac{0.5}{u_4} + \frac{1}{u_5}$$

Here, u_i ($i = 1, 2, 3, 4, 5$) are sub-intervals, while numbers in nominator denote membership degrees of u_i to A_i.

Step 3: Fuzzification of data:

For example, the data of 01.09.2014, (80824.72) is a value falling into sub-interval u_3. As the highest membership degree of the sub-interval u_3 is in A_3, this value is fuzzyfied as A_3. All of the values of data fuzzyfied in similar.

Step 4: Definition of fuzzy logic relationships:

For example, Group 1 defined as follow.

$A_1 \rightarrow A_1, A_1, A_1, A_1, A_1, A_1, A_1, A_1, A_1, A_1, A_1, A_1, A_1, A_1, A_1, A_1, A_1, A_1, A_1, A_2$

In a similar way, other groups are defined according to fuzzy relations.

Step 5: Establish Markov state transition matrix and transition process:

Transition probabilities matrix P is obtained by using from the fuzzy relationships in Step 4. Here, defining 5 states for each one of the fuzzy sets, 5×5 dimensional matrix is produced.

$$P_{ij} = \begin{bmatrix} \frac{21}{22} & \frac{1}{22} & 0 & 0 & 0 \\ \frac{2}{45} & \frac{34}{45} & \frac{9}{45} & 0 & 0 \\ 0 & \frac{10}{91} & \frac{69}{91} & \frac{12}{91} & 0 \\ 0 & 0 & \frac{12}{71} & \frac{57}{71} & \frac{2}{71} \\ 0 & 0 & 0 & \frac{2}{22} & \frac{20}{22} \end{bmatrix} \qquad (7)$$

Markov transition probabilities matrix given above will be used to create Markov transition probabilities. The related diagram is illustrated in Fig. (**2**).

Fig. (2). Transition Process.

Step 6: Calculation of forecasted values:

For example, the forecasted value for 02/09/2014 is calculated as follows:

$$F(02/09/2014) = \frac{10}{91} m_2 + \frac{69}{91} Y(01/09/2014) + \frac{12}{91} m_4 \qquad (8)$$

$$F(02/09/2014) = \frac{10}{91} 77359 + \frac{69}{91} 80824.72 + \frac{12}{91} 85395 \approx 81041.7987 \qquad (9)$$

Step 7: Adjusting forecasted values:

In this step, necessary adjustments are performed according to defined rules given in the previous section. For example, 17th observation of the data is employed to predict 18th observation. Corresponding fuzzified values are A_1 and A_2 for 17th and 18th observation, respectively. Thus, there is $A_2 \rightarrow A_1$ transition for 18th observation. According to the related transition process diagram, it is obvious that this is an accessible state. Therefore, adjustment given below is performed.

$$D_{t1} = -\left(\frac{l}{2}\right) = -4018.4 \qquad (10)$$

Let F and F' represent prediction and adjusted values, respectively. The adjusted value is calculated as follows:

$$F'(18) = F(18) - D_{t2} = 77463 - 4018.4 = 73444.6 \qquad (11)$$

All predictions are adjusted in a similar way.

In order to evaluate the obtained results for both predictions and forecasts, Mean

Absolute Percentage Error (MAPE) and Root Mean Square Error (RMSE) are employed. The related formulas are given below.

$$MAPE = \frac{1}{n} \sum_{i=1}^{n} \left| \frac{Forecasted\ value_i - Actual\ value_i}{Actual\ value_i} \right| \%100 \tag{12}$$

$$RMSE = \sqrt{\frac{\sum_{i=1}^{n} \left(Forecasted\ value_i - Actual\ value_i \right)^2}{n}} \tag{13}$$

where n is the number of observations.

Tsaur's [17] fuzzy time series forecasting model is applied to the time series for different numbers of fuzzy sets in order to reach better forecasting results. The number of fuzzy sets is changed from 5 to 15. For each number of fuzzy sets the fuzzy time series forecasting model is applied and MAPE and RMSE criteria values are calculated over both training and test sets. All obtained criteria values for each number of fuzzy sets are presented in Tables **1** and **2**. (Tables **1** and **2**) give the evaluation results for the training and the test sets, respectively.

Table 1. Evaluation results for the training set.

Number of fuzzy sets	MAPE	RMSE
5	1.1846	1644.4
6	1.4440	1808.2
7	1.1218	1457.7
8	1.1772	1430.5
9	1.1291	1328.4
10	1.1855	1330.7
11	1.2560	1343.1
12	1.1754	1219.2
13	1.0171	1082.3
14	1.0615	1078.7
15	1.0978	1072.6

According to Table **1**, the best training results are obtained for the numbers of fuzzy sets 13 and 15 in terms of MAPE and RMSE, respectively.

Table 2. Evaluation results for the test set.

Number of fuzzy sets	MAPE	RMSE
5	1.8699	1530.80
6	3.3828	2709.50
7	1.4025	1125.40
8	0.9635	779.30
9	0.8129	678.44
10	1.0071	810.22
11	1.0395	835.08
12	1.1420	909.56
13	1.2420	992.78
14	0.9522	773.26
15	0.8234	689.63

When Table **2** is examined, it is seen that Tsaur's [17] model produce the best forecasts when the number of fuzzy set is 15 in terms of the both performance measures.

For the aim of comparison, Chen's [18] fuzzy time series method which is a well-known method is also applied to the same real world time series. In a similar way, the number of fuzzy set is changed between 5 and 15. All obtained criteria values for the test set are summarized in Table **3**.

According to Table **3**, the forecasting model proposed by Chen [18] gives the best forecasts when the numbers of fuzzy set are 13 and 8 in terms of MAPE and RMSE, respectively. Because, in this case, Chen's method has the minimum forecasting error for both criteria. In order to compare the both methods, the best forecasting results obtained from both methods are presented in Table **4**.

Table 3. Forecasting results obtained from Chen's [18] method.

Number of fuzzy sets	MAPE	RMSE
5	1.5596	1594.3
6	1.8516	1822.4
7	1.3137	1321.2
8	1.4270	1030.8
9	1.4625	1482.9

(Table 3) contd.....

Number of fuzzy sets	MAPE	RMSE
10	1.1940	1243.6
11	1.3404	1331.2
12	1.2974	1320.0
13	1.0941	1149.2
14	1.2799	1298.7
15	1.2626	1273.7

According to Table **4**, it is clearly seen that Tsaur's [17] model produces more accurate forecasts than those obtained from Chen's [18] model in terms of the both criteria MAPE and RMSE. For IMKB100, Tsaur's [17] model gives better forecasts since Tsaur's [17] model has lowest value of RSME and MAPE. Forecasting performance of Tsaur's [17] model is also examined visually. The graph of forecasts obtained from Tsaur's [17] model and the corresponding observations can be seen in Fig. (**3**). In this Fig., observations and forecasts are represented by dashed and solid lines, respectively.

Table 4. The best forecasting results obtained from the both methods.

	MAPE	RMSE
Chen [18]	1.0941	1030.80
Tsaur [17]	0.8234	689.63

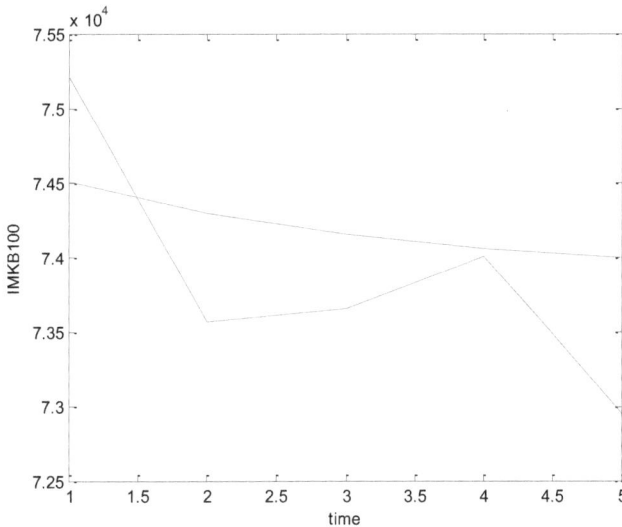

Fig. (3). Observations and forecasts obtained from Tsaur's [17] model.

CONCLUSION

Forecasting exchange rates is an important financial problem that is receiving increasing attention especially because of its difficulty and practical applications [20]. Forecasting exchange rates time series is a very difficult task since such time series also include uncertainty. Conventional time series methods cannot give satisfactory results for such time series. Therefore, fuzzy time series approaches should be employed to analyze such series according to the nature of the data.

There have been various fuzzy time series forecasting approaches in the literature since these models have proved their success in many applications from different fields. One of them is a forecasting method proposed by Tsaur [17]. His method is a first order fuzzy time series forecasting model in which Markov chain transition matrix is used to achieve fuzzy relationship groups. In other words, fuzzy logic relations are defined by using Markov chain transition matrix in his approach. Tsaur [17] showed that his method is superior to many methods available in the literature in terms of forecasting performance.

In this study, Tsaur's [17] fuzzy time series forecasting approach is explained. And, index 100 in stocks and bonds exchange market of İstanbul is forecasted by using Tsaur's [17] forecasting approach. For different numbers of fuzzy set, the performance of the Tsaur's [17] model is examined. Also, fuzzy time series model proposed by Chen [18] is applied to this real world time series for a comparison. As a result of the implementation, it is clearly seen that Tsaur's [17] fuzzy time series forecasting approach produces accurate forecasts for the real world data.

CONFLICT OF INTEREST

The authors (editor) declares no conflict of interest, financial or otherwise.

ACKNOWLEDGEMENTS

Declared none.

REFERENCES

[1]   C.H. Aladag, U. Yolcu, and E. Egrioglu, "A high order fuzzy time series forecasting model based on adaptive expectation and artificial neural networks", In: *Mathematics and Computers in Simulation* vol. 81. , 2010, pp. 875-882.

[2]   C.H. Aladag, and E. Egrioglu, "Advanced time series forecasting methods", In: *Advances in time series forecasting,* C.H. Aladag, E. Egrioglu, Eds., Bentham Science Publishers Ltd., 2012, pp. 3-10. eISBN: 978-1-60805-373-5, 2012.

[3]   C.H. Aladag, "Using artificial neural networks in fuzzy time series analysis", In: *Recent Developments*

and New Directions in Soft Computing. Studies in Fuzziness and Soft Computing 317., L.A. Zadeh, Ed., Springer International Publishing: Switzerland, 2014, pp. 443-451.
[http://dx.doi.org/10.1007/978-3-319-06323-2_28]

[4]   C.H. Aladag, "Using multiplicative neuron model to establish fuzzy logic relationships", *Expert Syst. Appl.,* vol. 40, no. 3, pp. 850-853, 2013.
[http://dx.doi.org/10.1016/j.eswa.2012.05.039]

[5]   C.H. Aladag, U. Yolcu, E. Egrioglu, and B. Eren, "Fuzzy lagged variable selection in fuzzy time series with genetic algorithms", *Appl. Soft Comput.,* vol. 22, pp. 465-473, 2014.
[http://dx.doi.org/10.1016/j.asoc.2014.03.028]

[6]   C.H. Aladag, and I.B. Turksen, "A novel membership value based performance measure", *J. Intell. Fuzzy Syst.,* vol. 28, no. 2, pp. 919-928, 2015.

[7]   C.H. Cheng, T.L. Chen, H.J. Teoh, and C.H. Chiang, "Fuzzy time series based on adaptive expectation model for TAIEX forecasting", *Expert Syst. Appl.,* vol. 34, pp. 1126-1132, 2008.
[http://dx.doi.org/10.1016/j.eswa.2006.12.021]

[8]   E. Egrioglu, C.H. Aladag, and U. Yolcu, "Fuzzy time series forecasting with a novel hybrid approach combining fuzzy c-means and neural networks", *Expert Syst. Appl.,* vol. 40, no. 3, pp. 854-857, 2013.
[http://dx.doi.org/10.1016/j.eswa.2012.05.040]

[9]   K. Huarng, "Effective lengths of intervals to improve forecasting in fuzzy time series", *Fuzzy Sets Syst.,* vol. 123, pp. 387-394, 2001.
[http://dx.doi.org/10.1016/S0165-0114(00)00057-9]

[10]   K. Huarng, "Heuristic models of fuzzy time series for forecasting", *Fuzzy Sets and Systems,* vol. 123, no. 3, pp. 369-386, 2001.
[http://dx.doi.org/10.1016/S0165-0114(00)00093-2]

[11]   K. Huarng, and H.K. Yu, "Ratio-based lengths of intervals to improve fuzzy time series forecasting", *IEEE Trans. Syst. Man Cybern. B Cybern.,* vol. 36, pp. 328-340, 2006.
[http://dx.doi.org/10.1109/TSMCB.2005.857093]

[12]   H.K. Yu, "Weighted fuzzy time series models for TAIEX forecasting", *Physica A,* vol. 624, pp. 609-624, 2005-a.
[http://dx.doi.org/10.1016/j.physa.2004.11.006]

[13]   H.K. Yu, "A refined fuzzy time series model for forecasting", *Physica A,* vol. 346, pp. 657-681, 2005-b.
[http://dx.doi.org/10.1016/j.physa.2004.07.024]

[14]   L.A. Zadeh, "Fuzzy sets", *Inf. Control,* vol. 8, pp. 338-353, 1965.
[http://dx.doi.org/10.1016/S0019-9958(65)90241-X]

[15]   Q. Song, and B.S. Chissom, "Fuzzy time series and its models", *Fuzzy Sets Syst.,* vol. 54, pp. 269-227, 1993.
[http://dx.doi.org/10.1016/0165-0114(93)90372-O]

[16]   Q. Song, and B.S. Chissom, "Forecasting enrollments with fuzzy time series-Part I", *Fuzzy Sets Syst.,* vol. 54, pp. 1-10, 1993.
[http://dx.doi.org/10.1016/0165-0114(93)90355-L]

[17]   R.C. Tsaur, "A fuzzy time series-Markov chain model with an application to forecast the exchange rate between the Taiwan and US dolar", *Int. J. Innov. Comput., Inf. Control,* vol. 8, pp. 1349-4198, 2011.

[18]   S.M. Chen, "Forecasting enrollments based on fuzzy time series", *Fuzzy Sets Syst.,* vol. 81, pp. 311-319, 1996.
[http://dx.doi.org/10.1016/0165-0114(95)00220-0]

[19]   U. Yolcu, E. Eğrioğlu, V.R. Uslu, M.A. Başaran, and C.H. Aladag, "A new approach for determining

the length of intervals for fuzzy time series", *Appl. Soft Comput.,* vol. 9, pp. 647-651, 2009.
[http://dx.doi.org/10.1016/j.asoc.2008.09.002]

[20]   W. Huang, K.K. Lai, Y. Nakomori, and S. Wang, "Forecasting foreign exchange rates with artificial neural networks", *A Review. International Journal of Information Technology & Decision Making,* vol. 3, no. 1, pp. 145-165, 2004.
[http://dx.doi.org/10.1142/S0219622004000969]

CHAPTER 7

A New High Order Multivariate Fuzzy Time Series Forecasting Model

Ufuk Yolcu[*]

Department of Econometrics, Faculty of Economics and Administrative Sciences, Giresun University, Giresun, Turkey

Abstract: In many disciplines, including uncertainty of data obtained from time series problems generates the needs to use fuzzy time series methods which do not need to check some strict assumptions of conventional time series methods. Although, there are many other well-known prediction methods in the fuzzy time series literature, most of them comprise of univariate methods and these methods may fail to satisfy to analysis of the data which contain multivariate relationships. In this study, the new multivariate fuzzy time series approach is proposed. The proposed approach uses fuzzy C-means method to determine the membership values in the fuzzification stage, and also this new multivariate approach makes use of single multiplicative neuron model artificial neural network for the identification of the multivariate fuzzy relations. In the identification of fuzzy relations stage, membership values are used to avoid the information loss. The proposed methods' performance has been assessed by applying it to different data sets.

Keywords: Artificial neural network, Forecasting,Fuzzy c-means, Membership degree, Multivariate fuzzy time series, Single multiplicative neuron model.

INTRODUCTION

Especially during the last few decades, fuzzy time series (FTS) methods have been searched by many researchers. There are two basic reasons for this. First of them, the data sets that encountered in various forecasting problem contain uncertainty and this kind of data should be analyzed by means of FTS methods. And also, second one, the analysis process of fuzzy time series approaches do not contain some strict assumptions on the contrary to conventional time series analysis procedures.

First fuzzy time series concept which based on Zadeh [1]'s fuzzy set theory was introduced by Song and Chissom [2]. Afterwards, Song and Chissom presented

---

[*] **Corresponding author Ufuk Yolcu:** Giresun University, Faculty of Economics and Administrative Sciences, Department of Econometrics, Giresun, 28200, Turkey; E-mail: varyansx@hotmail.com

two more studies which include fuzzy time series analysis methods [3, 4]. In fuzzy time series forecasting models, analysis process is consist of three basic step. These steps are generally named as fuzzification, identification of fuzzy relation and defuzzification, respectively. The effect of each steps on performance of forecasting models is inconvertible. For this reason, to be able to obtain more accurate forecasts, many researchers put forward various studies which search this impact.

For the fuzzification step, while the partition of universe of discourse has been utilized *via* sub-intervals in majority of studies, in the others some fuzzy clustering techniques have been used. In the partition of universe of discourse, two different approaches might be mentioned as the determining of equal intervals and the determining of dynamic intervals. Song and Chissom [2 - 4] and Chen [5, 6] determined these equal intervals, arbitrarily. Huarng [7] and Egrioglu and co-workers [8, 9] utilized some systematic approaches to determine them. In the subsequent studies, Kuo and co-workers [10, 11], Davari and co-workers [12], Park and co-workers [13] and Hsu and co-workers [14] used the particle swarm optimization, Chen and Chung [15] and Lee and co-workers [16, 17] used the genetic algorithm and Yolcu et. al. [18] utilized artificial bee colony algorithm to obtain the dynamic intervals. In addition, Cheng and co-workers [19], Li and co-workers [20] and Cagcag Yolcu [21] used fuzzy C-means (FCM).

For the identification of fuzzy relation, while Chen [5, 6] and various other researchers used the fuzzy logic group tables, Huarng and Yu [22], Aladag and co-workers [23], Egrioglu and co-workers [24 - 26], Yu and Huarng [27, 28] and Alpaslan and co-workers [29] used different types of artificial neural network (ANN) such as feed forward ANN (FF-ANN) and single multiplicative neuron model ANN (SMN-ANN).

In the defuzzification step, the centroid method is frequently used.

In some cases, a univariate forecasting model can be sufficient for a forecasting problem but in some other cases it may not be sufficient especially for the problem of forecast fuzzy time series which correlate with other fuzzy time series. In such circumstances, we need to constitute a bivariate or a multivariate model for forecasting the fuzzy time series. Hsu and co-workers [14] and Lee and co-workers [16, 17] used two factor (bivariate) fuzzy time series forecasting methods. Moreover, Jilani and Burney [30], Jilani and co-workers [31] and Egrioglu and co-workers [24] used multivariate fuzzy time series forecasting method. The membership degrees are taken into account in none of these studies. However, not taking into account the membership degrees leads to information loss and therefore the forecasting performance of method may be affected negatively. Yu and Huarng [28] who took in consideration this problem used an FFANN type, of which the input and output are the membership degrees in the determination of fuzzy relations. But in this study the membership degrees were defined, arbitrarily.

A multivariate high order fuzzy time series forecasting approach is presented in this paper. In this approach, the membership values have been obtained by using FCM technique. And also, SMN-ANN has been used for the determining of fuzzy relations.

In the rest of the paper, firstly, the fuzzy C-means clustering and SMN-ANN are briefly given. Secondly, fuzzy time series and its related definitions are presented. And then, the proposed model is introduced and the implementations of the proposed method and their results are presented for different data sets. Finally, conclusions and discussions are given.

RELATED METHODOLOGY

The Fuzzy C-Means (FCM) Clustering Method

Bezdek [32] introduced FCM clustering technique. This clustering technique is a widely used approach in the literature. In FCM algorithm, it is desired that least squared errors within groups is minimum. Let u_{ij} and v_i be the membership value and the cluster centre, respectively. And also, n represents the number of variables and c symbolise the number of clusters. The objective function to be minimized can be given as follows:

$$J_\beta(X,V,U) = \sum_{i=1}^{c} \sum_{j=1}^{n} u_{ij}^\beta d^2\left(x_j, v_i\right) \tag{1}$$

where β represents a constant which satisfies $\beta > 1$ and the fuzzy index. $d(x_j,v_i)$ $d(x_t,v_i)$ represents a measurement that measures the similarity between the center of the fuzzy cluster and the corresponding observation. J_β is minimized with respect to conditions given in (2).

$$0 \le u_{ij} \le 1 \qquad , \forall i,j$$

$$0 < \sum_{j=1}^{n} u_{ij} \le n \qquad , \forall i \tag{2}$$

$$\sum_{i=1}^{c} u_{ij} = 1 \qquad , \forall j$$

In this method, minimizing of least squared errors is performed by means of an iterative algorithm. In each iteration u_{ij} values and v_i are updated by using the Eq.3 and 4 formulas,

$$v_i = \frac{\sum_{j=1}^{n} u_{ij}^{\beta} x_j}{\sum_{j=1}^{n} u_{ij}^{\beta}} \tag{3}$$

$$u_{ij} = \frac{1}{\sum_{k=1}^{c} \left(\frac{d(x_j, v_i)}{d(x_j, v_k)}\right)^{2/(\beta-1)}} \tag{4}$$

Single Multiplicative Neuron Model Artificial Neural Network (SMN-ANN)

In neurons of feed-forward neural networks, the input signal is calculated based on addition function. Yadav and co-workers [33] proposed a single multiplicative neuron model. In the model, multiplication function is utilized in order to predict the neurons' input signal. Yadav and co-workers [33] showed that single multiplicative neuron model gives better forecasting performance for time series forecasting. Zhao and Yang [34] recommended the use of PSO instead of back propagation learning algorithm proposed by Yadav and co-workers [33] for the single multiplicative neuron model's training process. An example for single multiplicative neuron model of n inputs is given in Fig. (1).

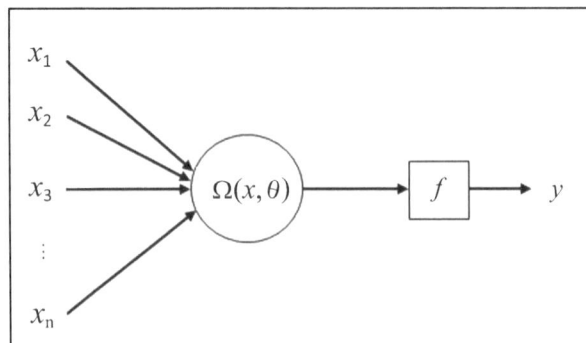

Fig. (1). The structure of single multiplicative neuron model.

This model has a single neuron and on the contrary to feed forward neural network, multiplication operation is performed. Function $\Omega(x,\Theta)$ is the product of the weighted inputs. The multiplicative neural model for n inputs is shown with Fig. (1) (x_i, $i = 1,2, \ldots, n$) has $n \times 2$ weights. Of these, n are the weights corresponding to the inputs (w_i, $i = 1,2, \ldots, n$) and n to the sides of the weights (b_i,

$i = 1,2, \ldots, n$). In the study logistic activation function is used given with Eq.5.

$$f(x) = \frac{1}{1+e^{-x}} \tag{5}$$

then, the neuron's net value is taken as follows.

$$net = \Omega(x, \theta) = \prod_{i=1}^{n}(w_i x_i + b_i) \tag{6}$$

Thereby, weight output can be gained as $y = f(net)$ by means of activation function.

The calculated fitness function in the multiplicative neuron model's training process with PSO can be used as a criterion as the sum of squares which is obtained from the output values difference for all learning samples and target values.

$$SSE = \sum_{i=1}^{n}(d_i - y_i)^2 \tag{7}$$

where d_i and y_i point respectively to the objective value and i^{th} learning sample's network output.

Fuzzy Time Series

In the literature, Song and Chissom [2 - 4] were proposed first FTS study. In contrast to conventional time series methods, FTS approaches do not require to be satisfied a variety of assumptions. Being able to work with even a very small data set can be counted as another advantages of FTS models. The basic definitions of FTS is represented as follows:

"Let U represents the universe of discourse, where $U = \{u_1, u_2, \ldots, u_n\}$. A fuzzy set A_i of U can be defined as given in Eq. 8,

$$A_i = \frac{f_{A_i}(u_1)}{u_1} + \frac{f_{A_i}(u_2)}{u_2} + \cdots + \frac{f_{A_i}(u_n)}{u_n} \tag{8}$$

where f_{A_i} is the membership function of the fuzzy set A_i and $f_{A_i}; U \rightarrow [0,1]$. In addition to $f_{A_i}(u_j), j = 1,2, \ldots, n$ denotes a generic element of fuzzy set $A_i; f_{A_i}(u_j)$ is the degree of belongingness of u_1 to A_i; $f_{A_i}(u_j) \in [0,1]$.

Definition 1. Fuzzy time series Let $Y(t)$ ($t = \ldots,0,1,2,\ldots$) a subset of real numbers, be the universe of discourse by which fuzzy sets $f_i(t)$ are defined. If $F(t)$ is a collection of $f_1(t), f_2(t), \ldots$ then $F(t)$ is called a fuzzy time series defined on $Y(t)$.

Definition 2. Fuzzy time series forecasting models.

- Assume that $F(t-1)$ causes $F(t)$, then the relation can be stated as $F(t) = F(t-1)*R(t, t-1)$ and this expression represents the fuzzy logical elationship between $F(t)$ and $F(t-1)$, where "*" is an operator. When $F(t-1) = A_i$ and $(t) = A_j$, The fuzzy relation between $F(t)$ and $F(t-1)$ can be expressed as $A_i \rightarrow A_j$ where the left-hand side of relation is made up of A_i (current state) and the right-hand side of the fuzzy logical relationship consists of A_j (next state) refers to. Furthermore, these fuzzy logical relationships can be adapted to establish grouped fuzzy relationships.
- If the lagged fuzzy time series $F(t-1), F(t-2), \ldots, (t-m)$ lead to $F(t)$, then fuzzy logical relationship which is called the mth order fuzzy time series forecasting model can be given as:

$$F(t-m), F(t-m+1), \cdots F(t-2), F(t-1) \rightarrow F(t) \tag{9}$$

Where "$F(t-m), F(t-m+1), \ldots F(t-2), F(t-1)$" refers to the current state and $F(t)$ refers to the next state.

- Let $F_1(t)$ and $F_2(t)$ be two fuzzy time series. Suppose that $F_1(t-1) \rightarrow A_i$, $F_2(t-1) \rightarrow B_k$, and $F_1(t) \rightarrow A_j$. A bivariate fuzzy logic relationship is defined as $A_i, B_k \rightarrow A_j$, where, for the bivariate fuzzy logical relationship, the left hand side is made up of A_i, B_k and the right hand side consist of A_j. Therefore, the first order bivariate fuzzy time series forecasting model can be stated as given in Eq. 10.

$$F_1(t-1), F_2(t-1) \rightarrow F_1(t) \tag{10}$$

Where while $F_1(t)$ is called as the main factor, $F_2(t)$ is called as the secondary factor fuzzy time series, ($t = \ldots,0,1,2,\ldots$).

- Let $F_1(t)$ and $F_2(t)$ be two fuzzy time series. If $F_2(t)$ is caused by ($F_1(t-1), F_2(t-1)$),($F_1(t-2), F_2(t-2)$),$\ldots,F_1(t-m), F_2(t-m)$ so Eq. 11 shows fuzzy logic relationship which is called the two-factors m^{th} order fuzzy time series forecasting model.

$$\left.\begin{array}{l}(F_1(t-m), F_2(t-m)), \\ \vdots \\ \vdots \\ (F_1(t-2), F_2(t-2)), \\ (F_1(t-1), F_2(t-1))\end{array}\right\} \rightarrow F_1(t) \qquad (11)$$

Where while $F_1(t)$ is called as the main factor, $F_2(t)$ is called as the secondary factor fuzzy time series.

- Let $F_1(t),F_2(t),\ldots,F_k(t)$ be m fuzzy time series. If $F_1(t)$ is caused by $(F_1(t-1),F_2(t-1),\ldots,F_k(t-1))$, $(F_1(t-2),F_2(t-2),\ldots,F_k(t-2))$, $(F_1(t-m),F_2(t-m),\ldots,F_k(t-m))$ then this fuzzy logical relationship is represented by Eq. 12.

$$\left.\begin{array}{l}(F_1(t-m), F_2(t-m), \cdots F_k(t-m)), \\ \vdots \\ \vdots \\ (F_1(t-2), F_2(t-2), \cdots F_k(t-2)), \\ (F_1(t-1), F_2(t-1), \cdots F_k(t-1))\end{array}\right\} \rightarrow F_1(t) \qquad (12)$$

and it is called as the k-factors m^{th} order fuzzy time series forecasting model, where $F_1(t)$ and $F_2(t),F_3(t),\ldots,F_k(t)$ are named fuzzy time series' main factor and the secondary factors of FTS, respectively ($t = \ldots,0,1,2,\ldots$)."

THE PROPOSED METHOD

The proposed method, in its analysis process, uses both FCM technique depending theory of fuzzy set and SMNM-ANN which has flexible calculation ability. The method which is presented in this paper has the advantage of being the first study in the literature regarding multivariate high-order fuzzy time series forecasting in which membership degrees are taken into account to define fuzzy relations by using SMNM_ANN. The membership values generated by FCM are processed by SMNM-ANN to determine fuzzy relations and crisp forecasts. The method's algorithm can be given step by step.

Step 1. Determine the model order and construct the lagged variable.

The model order d which is planned as high order and with k variables is defined. Then (d-1) lagged variables are constructed.

Step 2. Time series are fuzzified.

c which is called fuzzy sets number should be defined with the number of observations, n and the constraint $2 \leq c \leq n$. The data consisted of the observations of each $k{\times}d$ variables in the multivariate structure model is applied FCM and the fuzzy sets are determined. Then sets centers are calculated and the centres values of the fuzzy sets v_r, $r = 1,2,...,c$ are ordered ascending. These ordered sets are symbolized by L_r, $r = 1,2,...,c$. Finally, for each observation, the memberships degrees are obtained.

We illustrate the model with two variables for simplicity. Suppose that $X(t)$ is the time series to be forecasted and $Y(t)$ is the time series to be used as an explanatory variable. Let's assume that a second order model will be fitted and the number of fuzzy sets c is 3. After applying FCM, the centers of fuzzy sets and the membership for each observation, it means as a row, are presented in Table **1**. For example we can easily see from that table, for the third observation, that is the third row, the membership degree belong to the second fuzzy set (L_2) is 0,193983 (μ_{L_2}).

The number of inputs and outputs of FFANN which is used in the second stage of FTS method is equal to c. What is going to be the number of hidden layer neurons is decided by trial and error. Here, the point to be noted is that the number of neurons in the hidden layer should be chosen so that FFANN does not lose its ability of generalization.

Step 3. The fuzzy relations are determined by SMNM-ANN.

The architecture of a SMNM-ANN for the model with c sets is given Fig. (**2**).

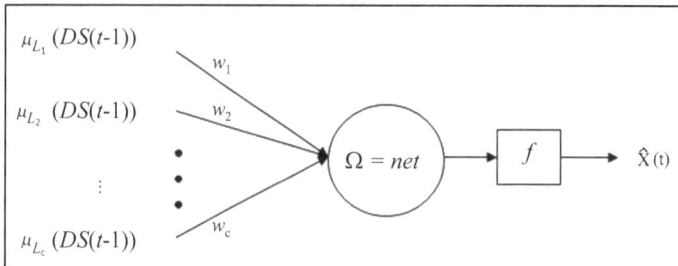

Fig. (2). The architecture of an SMNM-ANN with c fuzzy sets.

In Fig. (**2**) μ_{L_i} (DS(t)) represents the membership values for the data set at time *t*. The membership degrees in the left side of the Fig. are inputs of the SMNM-ANN, in the right side are the target values of SMNM-ANN. The outputs of the system are forecasts at time *t*. $w_1, w_2, \ldots, w_c$ are the weights of SMNM-ANN.

For time series with two variables, given in Table **1** and for the architecture of SMNM-ANN, presented in Fig. (**2**), the inputs and the targets of the network are as in Table **2**.

Table 1. An example for fuzzification stage.

				Set 1 centre (v_1)	Set 2 centre (v_2)	Set 3 centre (v_3)
$X(t)$				7346.214	7459.926	7524.701
$Y(t)$				7368.646	7470.834	7548.423
$X(t-1)$				7368.029	7487.07	7550.88
$Y(t-1)$				7383.522	7500.169	7590.13
				The membership degrees of observations		
t	$X(t)$	$Y(t)$	$X(t-1)$	$Y(t-1)$	Set 1 (L_1)	Set 2 (L_2)	Set 3 (L_3)
1	7552	7599	---	---			
2	7560	7593	7552	7599	0.017022	0.077107	0.905871
3	7487	7500	7560	7593	0.025555	0.193983	0.780462
4	7462	7472	7487	7500	0.000110	0.999631	0.000259
5	7515	7530	7462	7472	0.076097	0.680865	0.243038
6	7365	7372	7515	7530	0.261166	0.554542	0.184292
7	7360	7384	7365	7372	0.984459	0.011441	0.004101
8	7330	7352	7360	7384	0.986328	0.009835	0.003838

Table 2. The inputs and the targets of the FFANN.

Training Sample	*t*	Input1 $\mu_{L_i}(DS(t-1))$	Input 2 $\mu_{L_i}(DS(t-1))$	Input 3 $\mu_{L_i}(DS(t-1))$	Target $\hat{X}_t$
1	3	0.017022	0.077107	0.905871	7487
2	4	0.025555	0.193983	0.780462	7462
3	5	0.000110	0.999631	0.000259	7515
4	6	0.076097	0.680865	0.243038	7365
5	7	0.261166	0.554542	0.184292	7360
6	8	0.984459	0.011441	0.004101	7330

APPLICATIONS

The features of the parameters of all applications are specified as follows:

- The model order (d) is changed between 1 and 10. This means that each data set was analyzed for 10 different forecasting models.
- The number of fuzzy sets (c) is changed between 5 and 15.

The following sigmoid activation function seen in Eq. 13 is utilized in all layers of SMNM-ANN.

$$f(x) = (1 + exp(-x))^{-1} \tag{13}$$

the learning process of SMNM-ANN was performed by Levenberg-Marquardt learning algorithm.

The following criteria given in Eq. 14-16 were used to evaluate the results. These are the root of mean square error (RMSE), the mean absolute percentage error (MAPE) and the direction accuracy (DA).

$$RMSE = \sqrt{\frac{\sum_{t=1}^{n}(X_t - \hat{X}_t)^2}{n}} \tag{14}$$

$$MAPE = \frac{1}{n}\sum \left|\frac{X_t - \hat{X}_t}{X_t}\right| \tag{15}$$

$$DA = \frac{1}{n-1}\sum_{t=1}^{n-1}\begin{cases} 1, & (X_{t+1} - X_t)(\hat{X}_{t+1} - X_t) > 0 \\ 0, & o.w. \end{cases} \tag{16}$$

Firstly, we applied the proposed method to the data of Taiwan Futures Exchange (TAIFEX) observed in the period 03.08.1998-30.09.1998 which were given Table **3**. In this application, the dependent variable (main factor) is "TAIFEX" $X(t)$ and the secondary (explanatory variable) variable is Taiwan Stock Exchange "TAIEX" $Y(t)$. That is, the number of variables is $k = 2$. The performance criteria were calculated for the test data while the determination of fuzzy relations was being generated by the optimization of the training data. The last 16 observation of the time series was chosen as the test data.

The performance measures and the forecasts belonging to the best case of each method are summarized Table **4**.

When Table **4** is examined, it is clearly seen that the proposed multivariate method has better forecasting performance than the others. The forecasted values and the observations for the test data are presented in Fig. (**3**).

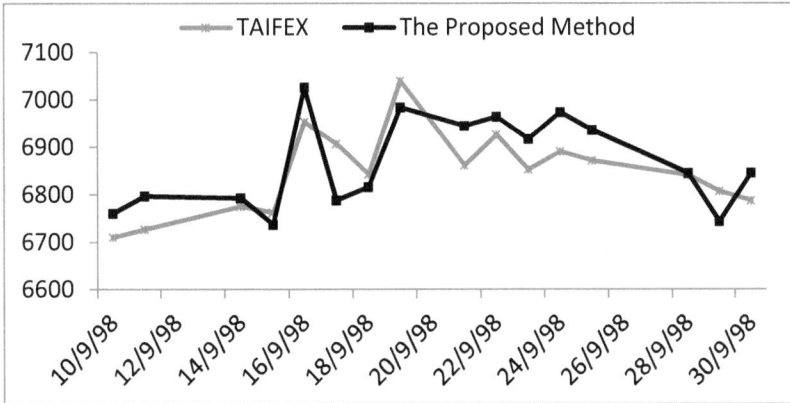

Fig. (3). The forecasted values and the observations for the test data of TAIFEX.

Table 3. The observations of "TAIFEX" and "TAIEX".

Date	TAIFEX	TAIEX	Date	TAIFEX	TAIEX	Date	TAIFEX	TAIEX
03.08.1998	7552.00	7599.00	24.08.1998	6955.00	6958.00	11.09.1998	6726.50	6842.00
04.08.1998	7560.00	7593.00	25.08.1998	6949.00	6908.00	14.09.1998	6774.55	6860.00
05.08.1998	7487.00	7500.00	26.08.1998	6790.00	6814.00	15.09.1998	6762.00	6858.00
06.08.1998	7462.00	7472.00	27.08.1998	6835.00	6813.00	16.09.1998	6952.75	6973.00
07.08.1998	7515.00	7530.00	28.08.1998	6695.00	6724.00	17.09.1998	6906.00	7001.00
10.08.1998	7365.00	7372.00	29.08.1998	6728.00	6736.00	18.09.1998	6842.00	6962.00
11.08.1998	7360.00	7384.00	31.08.1998	6566.00	6550.00	19.09.1998	7039.00	7150.00
12.08.1998	7330.00	7352.00	01.09.1998	6409.00	6335.00	21.09.1998	6861.00	7029.00
13.08.1998	7291.00	7363.00	02.09.1998	6430.00	6472.00	22.09.1998	6926.00	7034.00
14.08.1998	7.320.00	7348.00	03.09.1998	6200.00	6251.00	23.09.1998	6852.00	6962.00
15.08.1998	7320.00	7372.00	04.09.1998	6403.20	6463.00	24.09.1998	6890.00	6980.00
17.08.1998	7219.00	7274.00	05.09.1998	6697.50	6756.00	25.09.1998	6871.00	6980.00
18.08.1998	7220.00	7182.00	07.09.1998	6722.30	6801.00	28.09.1998	6840.00	6911.00
19.08.1998	7285.00	7293.00	08.09.1998	6859.40	6942.00	29.09.1998	6806.00	6885.00
20.08.1998	7274.00	7271.00	09.09.1998	6769.60	6895.00	30.09.1998	6787.00	6834.00
21.08.1998	7225.00	7213.00	10.09.1998	6709.75	6804.00			

Table 4. Observations and forecasts for the test data.

Date	TAIFEX	Lee and co-workers [16]	Lee and co-workers [17]	Hsu and co-workers [14]	The Proposed Method
10.09.1998	6709.75	6621.43	6917.40	6745.45	6759.58
11.09.1998	6726.50	6677.48	6852.23	6757.89	6796.10
14.09.1998	6774.55	6709.63	6805.71	6731.76	6792.25
15.09.1998	6762.00	6732.02	6762.37	6722.54	6736.31
16.09.1998	6952.75	6753.38	6793.06	6753.72	7025.71
17.09.1998	6906.00	6756.02	6784.40	6761.54	6787.19
18.09.1998	6842.00	6804.26	6970.74	6857.27	6814.98
19.09.1998	7039.00	6842.04	6977.22	6898.97	6982.86
21.09.1998	6861.00	6839.01	6874.46	6853.07	6944.02
22.09.1998	6926.00	6897.33	7126.05	6951.95	6963.83
23.09.1998	6852.00	6896.83	6862.49	6896.84	6916.39
24.09.1998	6890.00	6919.27	6944.36	6919.94	6972.79
25.09.1998	6871.00	6903.36	6831.88	6884.99	6935.23
28.09.1998	6840.00	6895.95	6843.24	6894.10	6843.96
29.09.1998	6806.00	6879.31	6858.45	6866.17	6742.45
30.09.1998	6787.00	6878.34	6825.64	6865.06	6844.71
	RMSE	93.49	102.96	80.02	62.48
	MAPE	1.09%	1.14%	0.87%	0.82%
	DA	53.33%	80.00%	73.33%	86.67%

Secondly, Taiwan Stock Exchange (TAIEX) data for the period 02.01.2004 and 31.12.2004 which was often used in many studies was analyzed by the proposed method. The last 45 observations have been used as the test data. The plot of time series the data is given (Fig. **4**).

In this application, "TAIEX" $X(t)$ was determined as main factor and the indexes' of "DOW JONES" $(Y_1(t))$ and "NASDAQ" $(Y_2(t))$ was determined as the secondary factors. So the number of variables in the system is to be 3. Evaluation of the forecasting performance was carried out over the test set of the data.

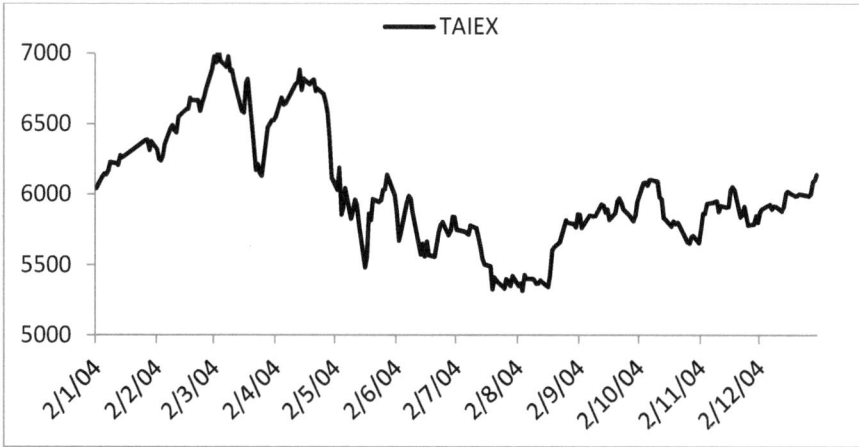

Fig. (4). The time series plot of "TAIEX" (Taiwan Stock Exchange).

The methods summarized in Table **5** were applied to the TAIEX data. The results on RMSE were obtained from the best case for all methods and shown in Table **5**.

Table 5. The results for the test data.

Methods	RMSE
Song and Chissom [1]	77.86
Chen [5] .	77.18
Chen [6]	71.98
Huarng and Yu [22]	63.57
Huarng and co-workers [35]	72.35
Yu and Huarng [28]	67.00
Aladag and co-workers [23]	63.01
Chen and Chen [36]	57.73
The proposed method	56.03

As you can see from Table **5**, the best performance was obtained from the proposed method. This situation is supported by Fig. (**5**) which shows apparently the closeness the fitted values to the observations.

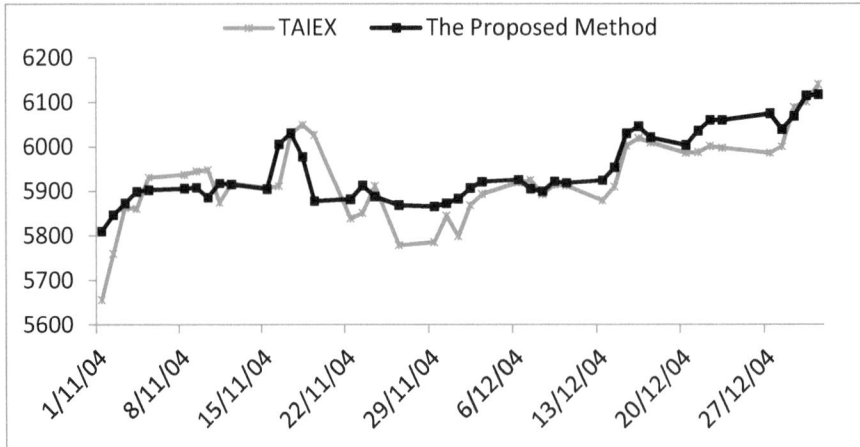

Fig. (5). The forecasted values and the observations for the test data of TAIEX.

CONCLUSIONS AND DISCUSSION

The partition of universe of discourse, which is a subjective approach, is used in almost every study in fuzzy time series literature. Then fuzzy clustering techniques has been began to use instead of it. Moreover, different ANN models have been begun to use for the identification fuzzy relations. In all of mentioned researches using ANN the inputs of the network are the index number of the fuzzified observations since the original observations are fuzzified by the index number of the fuzzy set with the highest membership degree. Therefore other membership degrees of that observation are ignored. This causes a loss of information. We propose to regard all of the membership degrees of observations. First of all we need to find all membership degrees by the way which is not arbitrary. We do this by using FCM to find memberships for each observation. After that, all memberships degrees of an observation are given to the network as inputs and the outputs are requested as membership degrees. In defuzzification stage we allow an approach which uses all membership degrees. Finally, in this study, we propose a multivariate fuzzy time series forecasting model with all these things.

The proposed method was supported by applying three different time series data. And the results were evaluated with those from other approaches previously used in literature. From these results we conclude that the proposed method ensures outstanding forecasting performance.

CONFLICT OF INTEREST

The author (editor) declares no conflict of interest, financial or otherwise.

ACKNOWLEDGEMENTS

Declared none.

REFERENCES

[1] L.A. Zadeh, "Fuzzy Sets", *Inf. Control,* vol. 8, pp. 338-353, 1965.
[http://dx.doi.org/10.1016/S0019-9958(65)90241-X]

[2] Q. Song, and B.S. Chissom, "Fuzzy time series and its models", *Fuzzy Sets Syst.,* vol. 54, pp. 269-277, 1993.
[http://dx.doi.org/10.1016/0165-0114(93)90372-O]

[3] Q. Song, and B.S. Chissom, "Forecasting enrollments with fuzzy time series- Part I", *Fuzzy Sets Syst.,* vol. 54, pp. 1-10, 1993.
[http://dx.doi.org/10.1016/0165-0114(93)90355-L]

[4] Q. Song, and B.S. Chissom, "Forecasting enrollments with fuzzy time series- Part II", *Fuzzy Sets Syst.,* vol. 62, pp. 1-8, 1994.
[http://dx.doi.org/10.1016/0165-0114(94)90067-1]

[5] S.M. Chen, "Forecasting enrolments based on fuzzy time-series", *Fuzzy Sets Syst.,* vol. 81, pp. 311-319, 1996.
[http://dx.doi.org/10.1016/0165-0114(95)00220-0]

[6] S.M. Chen, "Forecasting enrolments based on high order fuzzy time series", *Cybern. Syst.,* vol. 33, pp. 1-16, 2002.
[http://dx.doi.org/10.1080/019697202753306479]

[7] K. Huarng, "Effective length of intervals to improve forecasting in fuzzy time-series", *Fuzzy Sets Syst.,* vol. 123, pp. 387-394, 2001.
[http://dx.doi.org/10.1016/S0165-0114(00)00057-9]

[8] E. Egrioglu, C.H. Aladag, U. Yolcu, V.R. Uslu, and M.A. Basaran, "Finding an optimal interval length in high order fuzzy time series", *Expert Syst. Appl.,* vol. 37, pp. 5052-5055, 2010.
[http://dx.doi.org/10.1016/j.eswa.2009.12.006]

[9] E. Egrioglu, C.H. Aladag, M.A. Basaran, V.R. Uslu, and U. Yolcu, "A New Approach Based on the Optimization of the Length of Intervals in Fuzzy Time Series", *J. Intell. Fuzzy Syst.,* vol. 22, pp. 15-19, 2011.

[10] I-H. Kuo, S-J. Horng, T-W. Kao, T-L. Lin, C-L. Lee, and Y. Pan, "An improved method for forecasting enrollments based on fuzzy time series and particle swarm optimization", *Expert Syst. Appl.,* vol. 36, pp. 6108-6117, 2009.
[http://dx.doi.org/10.1016/j.eswa.2008.07.043]

[11] I-H. Kuo, S-J. Horng, Y-H. Chen, R-S. Run, T-W. Kao, R-J. Chen, J-L. Lai, and T-L. Lin, "Forecasting TAIFEX based on fuzzy time series and particle swarm optimization", *Expert Syst. Appl.,* vol. 37, pp. 1494-1502, 2010.
[http://dx.doi.org/10.1016/j.eswa.2009.06.102]

[12] S. Davari, M.H. Zarandi, and I.B. Turksen, "An Improved fuzzy time series forecasting model based on particle swarm intervalization", *Proc. The 28th North American Fuzzy Information Processing Society Annual Conferences (NAFIPS 2009)* Cincinnati, Ohio, USA
[http://dx.doi.org/10.1109/NAFIPS.2009.5156420]

[13] J-I. Park, D-J. Lee, C-K. Song, and M-G. Chun, "TAIFEX and KOSPI 200 forecasting based on two factors high order fuzzy time series and particle swarm optimization", *Expert Syst. Appl.,* vol. 37, pp. 959-967, 2010.
[http://dx.doi.org/10.1016/j.eswa.2009.05.081]

[14]   L-Y. Hsu, S-J. Horng, T-W. Kao, Y-H. Chen, R-S. Run, R-J. Chen, J-L. Lai, and I-H. Kuo, "Temperature prediction and TAIFEX forecasting based on fuzzy relationships and MTPSO techniques", *Expert Syst. Appl.,* vol. 37, pp. 2756-2770, 2010.
[http://dx.doi.org/10.1016/j.eswa.2009.09.015]

[15]   S.M. Chen, and N.Y. Chung, "Forecasting enrolments using high order fuzzy time series and genetic algorithms", *Int. J. Intell. Syst.,* vol. 21, pp. 485-501, 2006.
[http://dx.doi.org/10.1002/int.20145]

[16]   L.W. Lee, L.H. Wang, and S.M. Chen, "Temperature prediction and TAIFEX forecasting based on fuzzy logical relationships and genetic algorithms", *Expert Syst. Appl.,* vol. 33, pp. 539-550, 2007.
[http://dx.doi.org/10.1016/j.eswa.2006.05.015]

[17]   L.W. Lee, L.H. Wang, and S.M. Chen, "Temperature prediction and TAIFEX forecasting based on high-order fuzzy logical relationships and genetic simulated annealing techniques", *Expert Syst. Appl.,* vol. 34, pp. 328-336, 2008.
[http://dx.doi.org/10.1016/j.eswa.2006.09.007]

[18]   U. Yolcu, O. Cagcag, C.H. Aladag, and E. Egrioglu, "An enhanced fuzzy time series forecasting method based on artificial bee colony", *J. Intell. Fuzzy Syst.,* vol. 26, no. 6, pp. 2627-2637, 2014.

[19]   C-H. Cheng, G-W. Cheng, and J-W. Wang, "Multi-attribute fuzzy time series method based on fuzzy clustering", *Expert Syst. Appl.,* vol. 34, pp. 1235-1242, 2008.
[http://dx.doi.org/10.1016/j.eswa.2006.12.013]

[20]   S-T. Li, Y-C. Cheng, and S-Y. Lin, "A FCM-based deterministic forecasting model for fuzzy time series", *Comput. Math. Appl.,* vol. 56, pp. 3052-3063, 2008.
[http://dx.doi.org/10.1016/j.camwa.2008.07.033]

[21]   O. Cagcag Yolcu, ""A Hybrid Fuzzy Time Series Approach Based on Fuzzy Clustering and Artificial Neural Network with Single Multiplicative Neuron Model"", In: *Mathematical Problems in Engineering* vol. 2013. , 2013. Article ID 560472, 9 pages

[22]   K. Huarng, and H-K. Yu, "The application of neural networks to forecast fuzzy time series", *Physica A,* vol. 363, pp. 481-491, 2006.
[http://dx.doi.org/10.1016/j.physa.2005.08.014]

[23]   C.H. Aladag, M.A. Basaran, E. Egrioglu, U. Yolcu, and V.R. Uslu, "Forecasting in high order fuzzy time series by using neural networks to define fuzzy relations", *Expert Syst. Appl.,* vol. 36, pp. 4228-4231, 2009.
[http://dx.doi.org/10.1016/j.eswa.2008.04.001]

[24]   E. Egrioglu, C.H. Aladag, U. Yolcu, V.R. Uslu, and M.A. Basaran, "A new approach based on artificial neural networks for high order multivariate fuzzy time series", *Expert Syst. Appl.,* vol. 36, pp. 10589-10594, 2009.
[http://dx.doi.org/10.1016/j.eswa.2009.02.057]

[25]   E. Egrioglu, C.H. Aladag, U. Yolcu, M.A. Basaran, and V.R. Uslu, "A new hybrid approach based on SARIMA and partial high order bivariate fuzzy time series forecasting model", *Expert Syst. Appl.,* vol. 36, pp. 7424-7434, 2009.
[http://dx.doi.org/10.1016/j.eswa.2008.09.040]

[26]   E. Egrioglu, V.R. Uslu, U. Yolcu, M.A. Basaran, and C.H. Aladag, ""A new approach based on artificial neural networks for high order bivariate fuzzy time series"", In: *Applications of Soft Computing, AISC 58* Springer-Verlag Berlin Heidelberg, 2009, pp. 265-273.
[http://dx.doi.org/10.1007/978-3-540-89619-7_26]

[27]   H-K. Yu, and K. Huarng, "A bivariate fuzzy time series model to forecast TAIEX", *Expert Syst. Appl.,* vol. 34, pp. 2945-2952, 2008.
[http://dx.doi.org/10.1016/j.eswa.2007.05.016]

[28]   H-K. Yu, and K. Huarng, "A neural network- based fuzzy time series model to improve forecasting",

Expert Syst. Appl., vol. 37, pp. 3366-3372, 2010.
[http://dx.doi.org/10.1016/j.eswa.2009.10.013]

[29]   F. Alpaslan, O. Cagcag, C.H. Aladag, U. Yolcu, and E. Egrioglu, "A novel seasonal fuzzy time series method", *Hacet. J. Math. Stat.,* vol. 4, no. 3, pp. 375-385, 2012.

[30]   T.A. Jilani, and S.M. Burney, "Multivariate stochastic fuzzy forecasting models", *Expert Syst. Appl.,* vol. 35, no. 3, pp. 691-700, 2008.
[http://dx.doi.org/10.1016/j.eswa.2007.07.014]

[31]   T.A. Jilani, S.M. Burney, and C. Ardil, "Multivariate high order fuzzy time series forecasting for car road accidents", *International Journal of Computational Intelligence.,* vol. 4, no. 1, pp. 15-20, 2007.

[32]   J.C. Bezdek, *Pattern recognition with fuzzy objective function algorithms.* Plenum Press: New York, 1981.
[http://dx.doi.org/10.1007/978-1-4757-0450-1]

[33]   R.N. Yadav, P.K. Kalra, and J. John, "Time series prediction with single multiplicative neuron model", *Appl. Soft Comput.,* vol. 7, pp. 1157-1163, 2007.
[http://dx.doi.org/10.1016/j.asoc.2006.01.003]

[34]   L. Zhao, and Y. Yang, "PSO-based single multiplicative neuron model for time series prediction", *Expert Syst. Appl.,* vol. 36, pp. 2805-2812, 2009.
[http://dx.doi.org/10.1016/j.eswa.2008.01.061]

[35]   K. Huarng, H-K. Yu, and Y.W. Hsu, "A multivariate heuristic model for fuzzy time-series forecasting", *IEEE Trans. Syst. Man Cybern. B Cybern.,* vol. 37, no. 4, pp. 836-846, 2007.
[http://dx.doi.org/10.1109/TSMCB.2006.890303]

[36]   S.M. Chen, and C.D. Chen, "TAIEX forecasting based on fuzzy time series and fuzzy variation groups", *IEEE Trans. Fuzzy Syst.,* vol. 19, no. 1, pp. 1-12, 2011.
[http://dx.doi.org/10.1109/TFUZZ.2010.2073712]

CHAPTER 8

Fuzzy Functions Approach for Time Series Forecasting

Ali Z. Dalar[1,*], Erol Egrioglu[1], Ufuk Yolcu[2] and **Cagdas Hakan Aladag[3]**

[1] *Department of Statistics, Faculty of Arts and Sciences, Giresun University, Giresun, Turkey*

[2] *Department of Statistics, Faculty of Science, Ankara University, Ankara, Turkey*

[3] *Department of Mechanical and Industrial Engineering, University of Toronto, Toronto, Canada*

Abstract: After being proposed by Professor Lotfi A. Zadeh in 1965, fuzzy set theory has been used in many areas by researchers. Although fuzzy set theory has been used in many areas in the literature, implementation of fuzzy theories and techniques remains a difficult task, and it causes problems. One of these problems is to generate fuzzy if-then rules. These rules can be constituted by an inference of knowledge of experts, but human knowledge is generally incomplete. With the above problems in hand, in place of fuzzy rule base structures, fuzzy functions approach was proposed by Turksen in 2008. In literature, fuzzy inference systems have been used for forecasting problems. Classical fuzzy inference systems are based on the rules. As mentioned before, fuzzy functions are not based on the rules, and this is an advantage for them. Fuzzy functions approach was carried out to obtain forecasts by using simultaneous variables of other time series as covariates. In this chapter, type-1 fuzzy functions approach has been applied to obtain forecasts of Australian beer consumption and Turkey electricity consumption time series data. The lagged variables of elementary time series have been used as covariates. The performance of type-1 fuzzy functions approach has been evaluated against some recent methods in the literature.

Keywords: Artificial neural networks, Forecasting, Fuzzy c-means, Fuzzy time series, Type-1 fuzzy functions approach.

INTRODUCTION

Since late years, various alternative techniques have been used to forecast the time series. These techniques are usually based on artificial neural networks (ANNs) or fuzzy set theory. ANNs, when carried out to the time series, do not contain any approximations to uncertainty. The methods based on fuzzy set theory contain fuzzy approximation to uncertainty. Fuzzy set theory was proposed by Zadeh [1] and this theory has been used in many areas. Inference systems based on fuzzy set

* **Corresponding author Ali Z. Dalar:** Giresun University, Faculty of Arts and Sciences, Department of Statistics, 28100, Giresun, Turkey; E-mail: alizaferdalar@hotmail.com

theory which are similar to the inference mechanism of human brain, work with linguistic variables. The approaches for time series forecasting based on fuzzy set theory can be classified as fuzzy time series methods, fuzzy regression methods, fuzzy inference systems, and fuzzy functions approaches. Parameters of probabilistic models such as autoregressive models or traditional regression models are taken as fuzzy numbers in fuzzy regression methods. The purpose of these methods is to acquire more accurate point or interval predictions. These methods have not been often used in literature by the reason of using linear models and requiring computations on complex mathematical programming problems. Fuzzy time series (FTS) methods have wider area than fuzzy regression methods since it is easy to use FTS. Artificial intelligent systems and ANNs techniques have been easily used in FTS methods. One type of ANNs is multiplicative neuron model (MNM) that was proposed by Yadav *et al.* [2]. In last years, time series approaches based on MNM have been suggested by Zhao and Yang [3], Wu *et al.* [4], Wu *et al.* [5], Yolcu *et al.* [6], and Egrioglu *et al.* [7].

FTS approach was initially introduced by Song and Chissom [8], [9]. In recent years, MNM and membership values based FTS approaches have been proposed by Yu and Huarng [10], Egrioglu [11], Aladag *et al.* [12], Aladag *et al.* [13], Aladag [14], Cagcag Yolcu [15], and Egrioglu *et al.* [16]. Yolcu [17] proposed a high-order multivariate FTS forecasting model. These studies made a significant contribution to the literature. FTS approaches which work with membership values are like fuzzy functions approaches. Nevertheless, these types of FTS methods are first-order approaches, and some problems may occur when extending these approaches to high-order models. FTS methods have been used for time series forecast in the literature, and these methods do not contain any restrictions unlike classical time series methods. Membership values of FTS approaches are not taken into consideration in fuzzy inference system, and this is the main problem of these methods. Although many studies in recent years which take membership values into account, there is not yet sufficient literature in this subject. Many of FTS approaches are rule-based like classical fuzzy inference systems. Determining of the rules is an important problem in a fuzzy inference system, and also this is an important factor which affects performance of methods.

Fuzzy inference systems (FIS) can be designed from expert knowledge and learnt from data. FIS have been used for forecasting problem, but these methods have not been adequately used for time series forecasting problem. In literature, the most common used FIS for time series forecasting problem is adaptive neuro fuzzy inference system (ANFIS) that was proposed by Jang [18]. FIS are rule-based systems, and this is a disadvantage for them. For this reason, fuzzy functions approach was proposed. Inference system of fuzzy functions takes membership values into account. In place of fuzzy rule base structures, fuzzy functions approach was proposed by Turksen [19]. Fuzzy functions approach was proposed for fuzzy set theory based regression and clustering problems. Later on, fuzzy functions were developed by using different kinds of artificial intelligent systems and fuzzy sets (Celikyilmaz and Turksen [20 - 22], Turksen [23]). Beyhan and Alci [24] adjusted fuzzy functions to time series forecasting and used an embedded model. In Beyhan and Alci [24], the model was used as a linear ARX model, and lagged variables were identified by trial and error methods. A hybrid fuzzy functions approach was suggested by Zarandi *et al.* [25], and in this approach lagged variables were not used like in regression analysis. Aladag *et al.* [26] suggested a type-1 fuzzy functions approach. In the approach, inputs of the system are lagged variables of time series, and these variables are defined by binary particle swarm optimization.

In the literature, fuzzy functions approaches were applied to time series forecasting problem by using simultaneous other time series as covariates. However, it is well known that many time series can be explained with its or other time series' lagged variables and lagged variables should be used to obtain more accurate forecast. In this study, type-1 fuzzy functions approach which uses fuzzy C-means algorithm is implemented to two real world time series by using lagged variables of related time series for time series forecasting.

The paper is organized as follows. Section 2 starts with the introduction of type-1 fuzzy functions approach, and presents an algorithm for this approach. Section 3 presents obtained results from experimental studies. Finally, conclusions and discussions have been given in the last section.

TYPE-1 FUZZY FUNCTIONS APPROACH

Fuzzy functions approach, instead of rule-based FIS, was proposed by Turksen [19]. While a relation between input and output is constituted in rule-based FIS, a function is generated instead of a relation in fuzzy functions approach. There is no need to determine any rules in fuzzy functions approach.

The type-1 fuzzy functions approach algorithm (T1FF) is given below step by step.

Step 1. Inputs are lagged variables of time series. Matrix Z comprises of inputs and output of the system. Inputs and output of the system are clustered using FCM (Bezdek [27]) clustering method.

FCM clustering method can be applied by using the equations given below.

$$v_i = \frac{\sum_{k=1}^{n} (\mu_{ik})^{f_i} z_k}{\sum_{k=1}^{n} (\mu_{ik})^{f_i}} \; , \; i = 1, 2, ..., c \tag{1}$$

$$\mu_{ik} = \left[\sum_{j=1}^{c} \left(\frac{d(z_k, v_i)}{d(z_k, v_j)} \right)^{\frac{2}{f_i - 1}} \right]^{-1} \; , \; i = 1, 2, ..., c; \; k = 1, 2, ..., n \tag{2}$$

where f is degree of fuzziness, z_k is a vector whose elements are compose of kth row of Z, and μ_{ik} is degree of belongingness of kth observation to i^{th} cluster. $d(z, v)$ is Euclidian distance that is computed by using the equation (3).

$$d(z_k, v_i) = \| z_k - v_i \| \tag{3}$$

Step 2. Membership values of the input space are constituted as below.

$$\mu_{ik} = \left[\sum_{j=1}^{c} \left(\frac{d(x_k, v_i)}{d(x_k, v_j)} \right)^{\frac{2}{f_i - 1}} \right]^{-1} \; , \; i = 1, 2, ..., c; \; k = 1, 2, ..., n \tag{4}$$

where x is input matrix which is generated for lagged variables. If $\mu_{ik} \leq \alpha - cut$, then μ_{ik} value will be taken as zero.

Step 3. For each cluster i, membership values of each input data sample, μ_{ik} and original inputs are gathered together, and i^{th} fuzzy function is attained from predicting $Y^{(i)} = X^{(i)}\beta^{(i)} + \varepsilon^{(i)}$ multivariate regression model. When the number of the inputs is p, $X^{(i)}$ and $Y^{(i)}$ matrices are given below.

$$X^{(i)} = \begin{bmatrix} \mu_{i1} & x_{11} & \cdots & x_{p1} \\ \mu_{i2} & x_{12} & \cdots & x_{p2} \\ \vdots & \vdots & \ddots & \vdots \\ \mu_{in} & x_{1n} & \cdots & x_{pn} \end{bmatrix}, \; Y^{(i)} = \begin{bmatrix} y_1 \\ y_2 \\ \vdots \\ y_n \end{bmatrix} \tag{5}$$

Celikyilmaz and Turksen [22] used mathematical transformations of membership values. As they mentioned in their research, exponential and various logarithmic transformations of membership values can improve the performance of the system models.

In this study, for each cluster i by using membership values, μ_{ik} and/or their transformations such as μ_{ik}^2, $\exp(\mu_{ik})$, and $\ln((1 - \mu_{ik})/\mu_{ik})$ are taken into input matrix. After adding these transformations of membership values, matrix $X^{(i)}$ can be reconstituted as follow:

$$X^{(i)} = \begin{bmatrix} \mu_{i1} & \mu_{i1}^2 & \exp(\mu_{i1}) & \ln((1-\mu_{i1})/\mu_{i1}) & x_{11} & \cdots & x_{p1} \\ \mu_{i2} & \mu_{i2}^2 & \exp(\mu_{i2}) & \ln((1-\mu_{i2})/\mu_{i2}) & x_{12} & \cdots & x_{p2} \\ \vdots & \vdots & \vdots & \vdots & \vdots & \ddots & \vdots \\ \mu_{in} & \mu_{in}^2 & \exp(\mu_{in}) & \ln((1-\mu_{in})/\mu_{in}) & x_{1n} & \cdots & x_{pn} \end{bmatrix} \quad (6)$$

Step 4. Output values are computed by using the results obtained from fuzzy functions as follow:

$$\hat{y}_i = \frac{\sum_{i=1}^{c} \hat{y}_{ik} \mu_{ik}}{\sum_{i=1}^{c} \mu_{ik}}, \; k = 1,2,...,n \quad (7)$$

IMPLEMENTATION

T1FF approach was implemented to two real-world time series. These are Australian beer consumption and Turkey electricity consumption time series data.

Australian Beer Consumption Time Series

Australian beer consumption time series data (Janacek [28]) was quarterly observed between 1956-Q1 and 1994-Q1 whose graph is given in Fig. (**1**). The data consists of 148 observations in total, and the last 16 observations of the data were used as test set.

Australian beer consumption data was analyzed by using seasonal autoregressive integrated moving average (SARIMA), Winters' multiplicative exponential smoothing method (WMES), Yolcu *et al.* [29]'s linear and nonlinear artificial neural network model (L&NL-ANN), Aladag [14]'s multiplicative neuron model based fuzzy time series method (MNM-FTS), and T1FF approach. The number of lagged variables (m) is determined between 2 and 16, with an increment of 1. We experienced the number of fuzzy cluster (cn) between 3 and 10, with an increment of 1. α-*cut* is taken as 0 and 0.1.

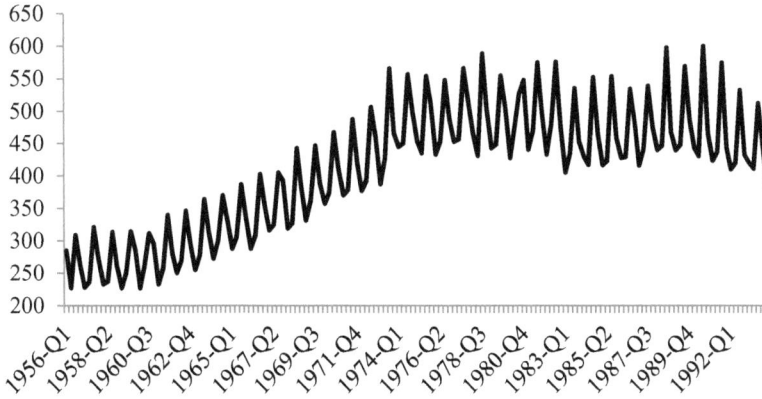

Fig. (1). The graph of Australian beer consumption time series data.

The results that obtained for Australian beer consumption time series can be seen in Table **1**.

Best model for SARIMA was attained as $SARIMA(0,1,1)(0,1,1)_4$. Linear trend component was utilized when WMES was carried out to the time series data. The best result for T1FF approach is obtained when *cn*, *m*, and α-*cut* are taken as 5, 8, and 0, respectively. Besides, the result is obtained when matrix $X^{(i)}$ is taken like equation (6). Root mean square error (RMSE) and mean absolute percentage error (MAPE) performance measures whose formulas are given in equation (8) and equation (9), respectively, are calculated for each method and values of these criteria are given in Table **1**.

Table 1. All obtained results for Australian beer consumption time series.

Date	Test Data	SARIMA	WMES	L&NL-ANN	MNM-FTS	T1FF
1989-Q1	430.50	452.72	453.91	449.92	437.50	446.20
1989-Q2	600.00	578.29	575.22	574.28	537.50	580.12
1989-Q3	464.50	487.70	502.32	481.47	437.50	483.04
1989-Q4	423.60	446.28	444.73	442.79	437.50	442.97
1990-Q1	437.00	456.77	459.66	445.12	437.50	444.74
1990-Q2	574.00	583.51	582.48	571.97	537.50	579.90
1990-Q3	443.00	492.13	508.64	472.76	487.50	468.01
1990-Q4	410.00	450.36	450.31	416.36	437.50	418.98
1991-Q1	420.00	461.01	465.40	428.63	437.50	431.60

(Table 1) contd.....

Date	Test Data	SARIMA	WMES	L&NL-ANN	MNM-FTS	T1FF
1991-Q2	532.00	588.96	589.74	559.89	562.50	559.41
1992-Q3	432.00	496.77	514.96	445.75	462.50	444.08
1992-Q4	420.00	454.64	455.89	390.25	412.50	394.99
1993-Q1	411.00	465.46	471.15	412.38	437.50	409.72
1993-Q2	512.00	594.71	597.00	533.19	537.50	525.60
1993-Q3	449.00	501.67	521.28	442.13	437.50	438.91
1993-Q4	382.00	459.17	461.46	405.08	412.50	409.07
	RMSE	47.0367	53.3295	18.7888	29.1381	17.3926
	MAPE	0.0949	0.1072	0.0357	0.0532	0.0345

$$RMSE = \sqrt{\frac{1}{T}\sum_{t=1}^{T}\left(y_t - \hat{y}_t\right)^2} \tag{8}$$

$$MAPE = \frac{1}{T}\sum_{t=1}^{T}\left|\frac{y_t - \hat{y}_t}{y_t}\right| \tag{9}$$

According to Table **1**, the most accurate forecasts are attained in terms of RMSE and MAPE criteria when T1FF approach is employed. The forecasting performance of T1FF approach can be seen visually, the graph of the real observations and the forecasts obtained from the approach for the test set, in Fig. (**2**). According to this graph, it is clearly seen that the forecasts obtained from T1FF approach are very accurate.

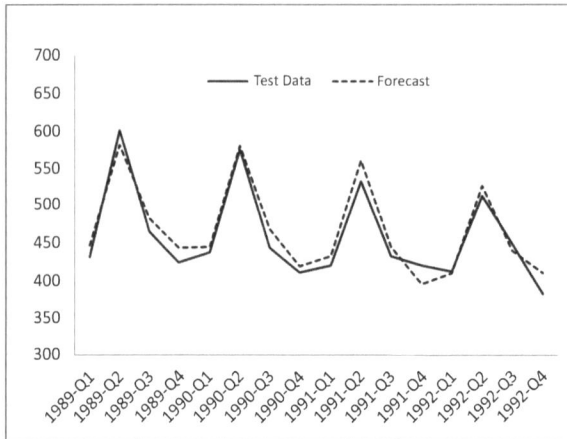

Fig. (2). The graph of the real observations and the forecasts obtained from the T1FF approach.

Turkey Electricity Consumption Time Series

Turkey electricity consumption data had been monthly observed between January, 2002 and December, 2013 whose given in Fig. (**3**). The data consists of 144 observations in total, and the last 12 observations of the data were used as test set.

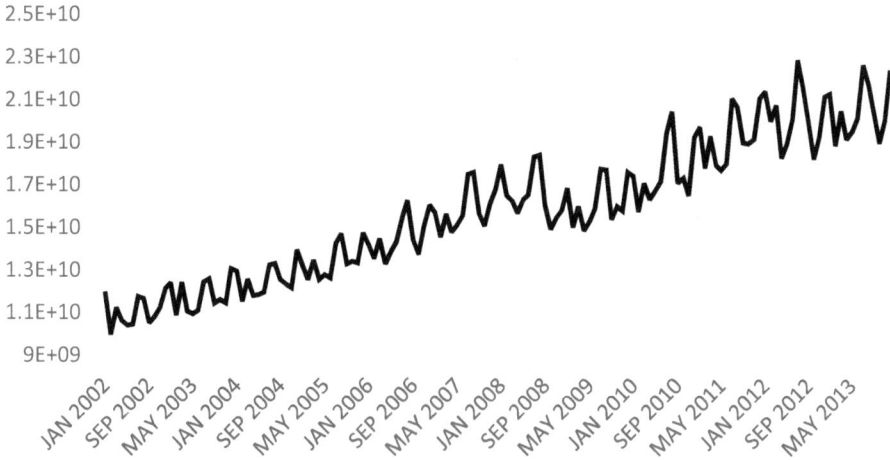

Fig. (3). The graph of Turkey electricity consumption time series data.

Turkey electricity consumption data was analyzed by using multilayer perceptron artificial neural network (MLP-ANN), seasonal autoregressive integrated moving average (SARIMA), multiplicative neuron model artificial neural network (MNM-ANN), Yolcu *et al.* [29] linear and nonlinear artificial neural network model (L&NL-ANN), and T1FF approach. The number of lagged variables (m) is determined between 2 and 16, with an increment of 1. We experienced the number of fuzzy cluster (cn) between 3 and 10, with an increment of 1. α-*cut* is taken as 0 and 0.1.

The results that obtained for Turkey electricity consumption time series can be seen in Table **2**. The best model for SARIMA was attained as SARIMA $(0,1,1)(0,1,1)_{12}$. The best result for T1FF approach is obtained when cn, m, and α-*cut* are taken as 4, 16, and 0.1, respectively. And besides, the result is obtained when matrix $X^{(i)}$ is taken like equation (6) except *ln* transformation. RMSE and MAPE performance measures are calculated for each method, and values of these criteria are given in Table **2**.

According to Table **2**, the most accurate forecasts are attained in terms of RMSE and MAPE criteria when T1FF approach is employed.

Table 2. All obtained results for Turkey electricity consumption time series.

Test Data	MLP-ANN	SARIMA	MNM-ANN	L&NL-ANN	T1FF
21275408487	21504260392	21690314907	21929976335	22005935580	21666285710
18841712637	21407932217	19879748318	19674336447	21273737442	20685602131
20463933683	19934741061	20679491118	20759353666	20386359964	19996267563
19139248871	17339157313	18616749529	18177421217	19068902087	17987700997
19511728912	19132107582	18166751109	19502618441	19337714416	19394021620
20132602347	20521392811	19476718063	19961628854	20493224715	20584842565
22648523194	21559624073	22996373189	21466603203	22493619106	22702026988
21698207982	21608712088	22602364226	20970484983	22642925619	21714016063
20358717408	20450417618	19372692274	20819409088	20387962879	19965826638
18964661109	19608790983	19145368564	18842139533	19384311943	18624614607
20061232838	20798229864	18281990717	20341486661	20668793095	20140094911
22405662577	21396071971	21319220705	20523920152	22329629382	22138069735
RMSE	1065870606	917321409	813259007	820978567	687192130
MAPE	0.039842	0.038842	0.030115	0.025444	0.023662

The forecasting performance of T1FF approach can be seen visually, the graph of the real observations and the forecasts obtained from the approach for the test set, in Fig. (**4**). According to this graph, it is clearly seen that the forecasts obtained from T1FF approach are accurate.

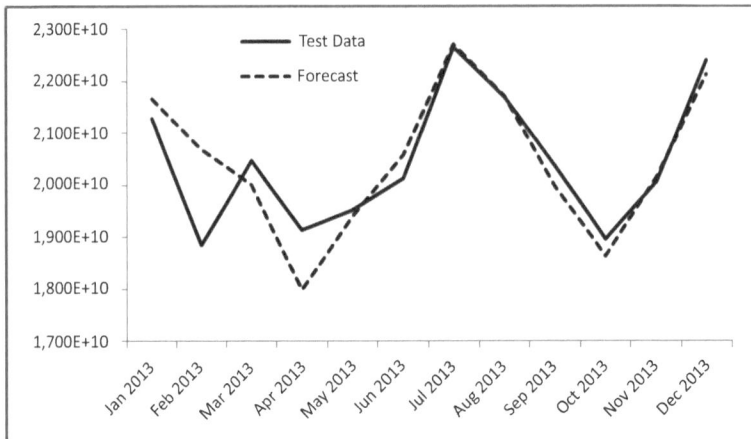

Fig. (4). The graph of the real observations and the forecasts obtained from the T1FF approach.

CONCLUSIONS

The most important feature of fuzzy functions is that in inference system of fuzzy functions are not needed the rules. In this study, T1FF approach is implemented to time series by using lagged variables. In the implementation, T1FF approach was applied to two real-world time series. The lagged variables of elementary time series have been used as covariates. The performance of T1FF approach has been compared with some recent methods such as ANNs and FTS methods available in the literature. As a consequence of the comparison, it was shown that T1FF approach produces the best forecasts in terms of RMSE and MAPE performance measures.

CONFLICT OF INTEREST

The authors (editor) declares no conflict of interest, financial or otherwise.

ACKNOWLEDGEMENTS

Declared none.

REFERENCES

[1]   L.A. Zadeh, "Fuzzy Sets", *Inf. Control,* vol. 8, no. 3, pp. 338-353, 1965.
       [http://dx.doi.org/10.1016/S0019-9958(65)90241-X]

[2]   R.N. Yadav, P.K. Kalra, and J. John, "Time series prediction with single multiplicative neuron model",
       Appl. Soft Comput., vol. 7, pp. 1157-1163, 2007.
       [http://dx.doi.org/10.1016/j.asoc.2006.01.003]

[3]   L. Zhao, and Y. Yang, "PSO-based single multiplicative neuron model for time series prediction",
       Expert Systems with Applications, vol. 36, no. 2, pp. 2805-2812, 2009.

[4]   X. Wu, Y. Tan, Y. Wang, and Y. Xiao, "Long-term prediction of time series with iterative extended
       Kalman Filter trained single multiplicative neuron model", *J. Comput. Inf. Syst.,* vol. 8, no. 14, pp.
       5933-5940, 2012.

[5]   X. Wu, J. Mao, Z. Du, and Y. Chang, "Online training algorithms based single multiplicative neuron
       model for energy consumption forecasting", *Energy,* vol. 59, pp. 126-132, 2013.
       [http://dx.doi.org/10.1016/j.energy.2013.06.068]

[6]   U. Yolcu, C.H. Aladag, E. Egrioglu, and V.R. Uslu, "Time series forecasting with a novel fuzzy time
       series approach: an example for Istanbul stock market", *J. Stat. Comput. Simul.,* vol. 83, no. 4, pp.
       599-612, 2013.
       [http://dx.doi.org/10.1080/00949655.2011.630000]

[7]   E. Egrioglu, U. Yolcu, C.H. Aladag, and E. Bas, "Recurrent multiplicative neuron model artificial
       neural network for non-linear time series forecasting", *Procedia Soc. Behav. Sci.,* vol. 109, pp. 1094-
       1100, 2014.
       [http://dx.doi.org/10.1016/j.sbspro.2013.12.593]

[8]   Q. Song, and B.S. Chissom, "Fuzzy time series and its models", *Fuzzy Sets Syst.,* vol. 54, pp. 269-277,
       1993.
       [http://dx.doi.org/10.1016/0165-0114(93)90372-O]

[9]   Q. Song, and B.S. Chissom, "Forecasting enrollments with fuzzy time series-Part I", *Fuzzy Sets Syst.,*

vol. 54, pp. 1-10, 1993.
[http://dx.doi.org/10.1016/0165-0114(93)90355-L]

[10] T.H. Yu, and K.H. Huarng, "A neural network-based fuzzy time series model to improve forecasting", *Expert Syst. Appl.,* vol. 37, pp. 3366-3372, 2010.
[http://dx.doi.org/10.1016/j.eswa.2009.10.013]

[11] E. Egrioglu, "A new time invariant fuzzy time series forecasting method based on genetic algorithm", *Advances in Fuzzy Systems,* vol. 2012, 2012. Article ID 785709.
[http://dx.doi.org/10.1155/2012/785709]

[12] C.H. Aladag, U. Yolcu, E. Egrioglu, and A.Z. Dalar, "A new time invariant fuzzy time series forecasting method based on particle swarm optimization", *Appl. Soft Comput.,* vol. 12, pp. 3291-3299, 2012.
[http://dx.doi.org/10.1016/j.asoc.2012.05.002]

[13] S. Aladag, C.H. Aladag, T. Mentes, and E. Egrioglu, "A new seasonal fuzzy time series method based on the multiplicative neuron model and SARIMA", *Hacet. J. Math. Stat.,* vol. 41, no. 3, pp. 337-345, 2012.

[14] C.H. Aladag, "Using multiplicative neuron model to establish fuzzy logic relationships", *Expert Syst. Appl.,* vol. 40, no. 3, pp. 15-, 850-853, 2013.
[http://dx.doi.org/10.1016/j.eswa.2012.05.039]

[15] O. Cagcag Yolcu, "A hybrid fuzzy time series approach based on fuzzy clustering and artificial neural network with single multiplicative neuron model", *Mathematical Problems in Engineering.,* vol. 2013, 2013. Article ID 560472.
[http://dx.doi.org/10.1155/2013/560472]

[16] E. Egrioglu, C.H. Aladag, and U. Yolcu, "Fuzzy time series method based on multiplicative neruin model and membership values", *American Journal of Intelligent Systems,* vol. 3, no. 1, pp. 33-39, 2013.
[http://dx.doi.org/10.5923/j.ajis.20130301.05]

[17] U. Yolcu, "The forecasting of Istanbul Stock Market with a high order multivariate fuzzy time series forecasting model", *Turkish Journal of Fuzzy Systems,* vol. 3, no. 2, pp. 118-135, 2012.

[18] J.S. Jang, "ANFIS: Adaptive network based fuzzy inference system", *IEEE Trans. Syst. Man Cybern.,* vol. 23, no. 3, pp. 665-685, 1993.
[http://dx.doi.org/10.1109/21.256541]

[19] I.B. Turksen, "Fuzzy function with LSE", *Appl. Soft Comput.,* vol. 8, pp. 1178-1188, 2008.
[http://dx.doi.org/10.1016/j.asoc.2007.12.004]

[20] A. Celikyilmaz, and I.B. Turksen, "Enhanced fuzzy system models with improved fuzzy clustering algorithm", *IEEE Trans. Fuzzy Syst.,* vol. 16, no. 3, pp. 779-794, 2008.
[http://dx.doi.org/10.1109/TFUZZ.2007.905919]

[21] A. Celikyilmaz, and I.B. Turksen, "Uncertainty modeling of improved fuzzy functions with evolutionary systems", *IEEE Trans. Syst. Man Cybern.,* vol. 38, no. 4, pp. 1098-1110, 2008.
[http://dx.doi.org/10.1109/TSMCB.2008.924587]

[22] A. Celikyilmaz, and I.B. Turksen, *Modeling uncertainty with fuzzy logic, Studies in Fuzziness and Soft Computing.* Springer, 2009, p. 240.
[http://dx.doi.org/10.1007/978-3-540-89924-2]

[23] I.B. Turksen, "Fuzzy system models", In: *Encyclopedia of Complexity and Systems Science,* 2009, pp. 4080-4094.

[24] S. Beyhan, and M. Alci, "Fuzzy functions based ARX model and new fuzzy basis function models for nonlinear system identification", *Appl. Soft Comput.,* vol. 10, pp. 439-444, 2010.
[http://dx.doi.org/10.1016/j.asoc.2009.08.015]

[25]   M.H. Zarandi, M. Zarinbal, N. Ghanbari, and I.B. Turksen, "A new fuzzy functions model tuned by hybridizing imperialist competitive algorithm and simulated annealing. Application: Stock price prediction", *Inf. Sci.,* vol. 222, no. 10, pp. 213-228, 2013.
[http://dx.doi.org/10.1016/j.ins.2012.08.002]

[26]   C.H. Aladag, U. Yolcu, E. Egrioglu, and I.B. Turksen, "Type-1 fuzzy time series function method based on binary particle swarm optimisation", *International Journal of Data Analysis Techniques and Strategies,* vol. 8, no. 1, pp. 02-13, 2016.
[http://dx.doi.org/10.1504/IJDATS.2016.075970]

[27]   J.C. Bezdek, *Pattern recognition with fuzzy objective function algorithms.* Plenum Press: New York, 1981.
[http://dx.doi.org/10.1007/978-1-4757-0450-1]

[28]   G.J. Janacek, *Practical time series.* Oxford University Press: New York, 2001.

[29]   U. Yolcu, E. Egrioglu, and C.H. Aladag, "A new linear & nonlinear artificial neural network model for time series forecasting", *Decision Support System Journals,* vol. 54, pp. 1340-1347, 2013.
[http://dx.doi.org/10.1016/j.dss.2012.12.006]

CHAPTER 9

Recurrent ANFIS for Time Series Forecasting

Busenur Sarıca[1,*], Erol Eğrioğlu[2] and Barış Aşıkgil[3]

[1] *Department of Statistics, Faculty of Arts and Sciences, Marmara University, İstanbul, Turkey*

[2] *Department of Statistics, Faculty of Arts and Sciences, Giresun University, Giresun, Turkey*

[3] *Department of Statistics, Faculty of Science and Letters, Mimar Sinan Fine Arts University, Istanbul, Turkey*

Abstract: A few recurrent ANFIS approaches were proposed in the literature. Two main types of recurrences are possible in ANFIS architecture. Feedback can be made for input layer or right sides of Sugeno-type rules. In this study, a new type recurrent ANFIS is proposed for forecasting. Feedback mechanism is embedded to ANFIS by using squares of error terms as inputs in right sides of Sugeno-type fuzzy rules. The training of the proposed ANFIS is made by using particle swarm optimization technique. The proposed method was tested on some real world time series data and it is compared with some alternative forecasting methods in the literature. It was shown that the proposed method has the best forecasting performance.

Keywords: ANFIS, Fuzzy C-Means, Fuzzy Inference System, Particle Swarm Optimization, Recurrent Networks.

INTRODUCTION

In the literature, fuzzy inference systems have been used for time series forecasting. Adaptive networks fuzzy inference systems (ANFIS) can produce accurate forecasts for lots of real-world time series. In processes of fuzzification and determining rules of ANFIS, some modifications for forecasting purpose have been made in the literature. In many forecasting applications, lagged variables were used as inputs of ANFIS. As a consequent of this, a nonlinear autoregressive model was obtained from ANFIS architecture in forecasting applications. The forecasting systems need lagged error variables for many real-world time series. Because of this, a few recurrent ANFIS methods were proposed in the literature. Zhang and Morris [1] used lagged errors as inputs of ANFIS, their method defined membership functions for the lagged error variables and the rules were determined by using independent variables. Zhang and Morris [1] method is suit-

---

* **Corresponding author Busenur Sarıca:** Statistics Department, Faculty of Arts and Science, Marmara University, İstanbul, Turkey; E-mail: busenur.sarica@marmara.edu.tr

Cagdas Hakan Aladag (Ed.)

able for regression problem and it was not organized for time series forecasting problem. Tamura *et al.* [2] proposed a recurrent ANFIS. In their method, one step lagged error variable and output of ANFIS was used as inputs of network. Yu and Tang [3] used lagged variables of error and outputs as inputs in their ANFIS method. Kasuan *et al.* [4] proposed a recurrent ANFIS and their method employees second order lagged error variable as input. Mahmud and Meesad [5] proposed a recurrent ANFIS with momentum unit. Momentum unit provides a differenced time series as input of network. Order of difference is order of momentum unit in Mahmut and Meesad's method. When these recurrent ANFIS are examined, it is understood that error variables and outputs of ANFIS were used as inputs of the recurrent ANFIS. When a recurrence embedded to the ANFIS, the number of inputs and so number of parameters are increased. This is an important disadvantage. Moreover, square of error variables can be useful for stock exchange time series but they were not used as inputs in ANFIS so far.

In this study, a new type recurrent ANFIS is proposed and it is called recurrent ANFIS (RANFIS). In the second section, RANFIS is introduced and its details are given. The application results are given in the third section. In the fourth section, the conclusions are given.

RECURRENT ADAPTIVE NETWORK FUZZY INFERENCE SYSTEMS

In the proposed method, squares of error variables are used as inputs in right sides of the rules but these inputs are not used to calculate power of the rules and there is no need to define membership functions for the squared error variables. Because of this, less number of parameters are used in the network than other recurrent ANFIS methods. In Fig. (**1**), an architecture example for RANFIS with two inputs and one step lagged squared error is given.

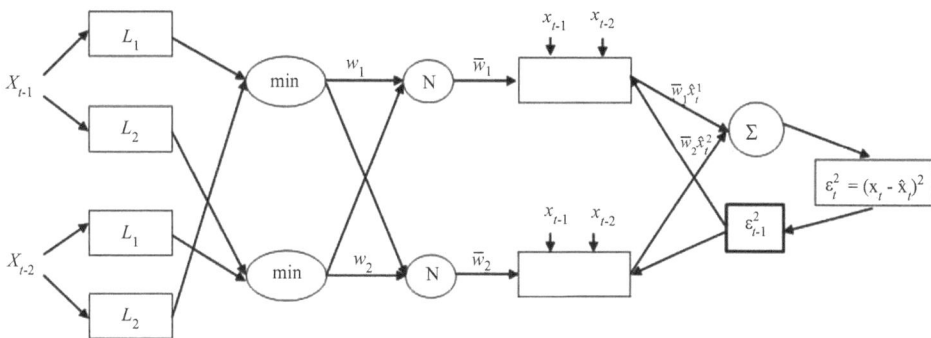

Fig. (1). RANFIS with two inputs and one step lagged squared error.

RANFIS are introduced with three algorithms. The first algorithm explains how to compute an output of RANFIS. The second algorithm is about training of RANFIS by using particle swarm optimization [6]. The third algorithm is the main algorithm for the RANFIS.

Algorithm 1: Calculation of an output from RANFIS

Step 1. Membership values (μ_L) are calculated according to fuzzy clusters (L_j) which were determined by fuzzy c-means method.

$\mu_L(x_{t-j})$: The membership value for j^{th} lagged variable to i^{th} fuzzy set

Step 2. The rule fire strengths are calculated for two rules (with two inputs) as follow:

$$w_1 = \min(\mu_{L_1}(x_{t-1}), \mu_{L_2}(x_{t-2})) \tag{1}$$

$$w_2 = \min(\mu_{L_2}(x_{t-1}), \mu_{L_1}(x_{t-2})) \tag{2}$$

Step 3. The normalized rule fire strengths are calculated as follow:

$$\bar{w}_i = \frac{w_i}{w_1 + w_2}, \quad i = 1, 2 \tag{3}$$

Step 4. According to consequent parameter values (p_i, q_i, m_i, r_i) and normalized rule strength, outputs of every rule are calculated as follow:

$$\bar{w}_i \hat{x}_t^i = \bar{w}_i(p_i x_{t-1} + q_i x_{t-2} + m_i \varepsilon_{t-1}^2 + r_i) \tag{4}$$

One step lagged square of the network error is taken as zero for only $t = 1$ and the others are calculated in Step 5.

Step 5. It is computed that combine output of all rule outputs as follow:

$$\hat{x}_t = \sum_i \bar{w}_i \hat{x}_t^i \tag{5}$$

Square of the network error is calculated as follow:

$$\varepsilon_t^2 = (x_t - \hat{x}_t)^2 \tag{6}$$

Algorithm 2. Training of RANFIS with Particle Swarm Optimization

Step 1. Positions of each kth (k = 1,2, …, pn) particles' positions are randomly determined and kept in a vector X_k given as follows:

$$X_k = \{x_{k,1}, x_{k,2}, …, x_{k,d}\}, \qquad k = 1, 2, …pn \qquad (7)$$

where $x_{k,i}$ (i=1,2,…,d) represents i^{th} position of kth particle. pn and d represents the number of particles in a swarm and positions, respectively. The positions of a particle are values for consequent part parameters.

Step 2. Velocities are randomly determined and stored in a vector V_k given below.

$$V_k = \{v_{k,1}, v_{k,2}, …, v_{k,d}\}, \qquad k = 1, 2, …pn \qquad (8)$$

Step 3. According to the evaluation function, *Pbest* and *Gbest* particles given in (9) and (10), respectively, are determined.

$$Pbest_k = \{pb_{k,1}, pb_{k,2}, …, pb_{k,d}\}, \qquad k = 1, 2, …pn \qquad (9)$$

$$Gbest = \{p_{g,1}, p_{g,2}, …, p_{g,d}\} \qquad (10)$$

where $Pbest_k$ is a vector stores the positions corresponding to the kth particle's best individual performance, and *Gbest* represents the best particle, which has the best evaluation function value, found so far. The evaluation function is selected as root of mean square error (RMSE) value which is computed for training data. RMSE can be calculated by using formula given in (11).

$$RMSE = \sqrt{\frac{1}{n}\sum_{t=1}^{n}(x_t - \hat{x}_t)^2} \qquad (11)$$

Where "n" is length of training data, $\hat{x}_t$ is forecast for t^{th} observation of time series and it is calculated by using Algorithm 1.

Step 4. Values of velocities and positions are updated by using the formulas given in (12) and (13), respectively.

$$v_{i,d}^{t+1} = \begin{bmatrix} w \times v_{i,d}^t + c_1 \times rand1 \times (p_{i,d} - x_{i,d}^t) + \\ c_2 \times rand2 \times (p_{g,d} - x_{i,d}^t) \end{bmatrix} \qquad (12)$$

$$x_{i,d}^{t+1} = x_{i,d}^{t} + v_{i,d}^{t+1} \tag{13}$$

where *rand1*, *rand2* are random values from the interval [0,1] and *c1*, *c2* are the acceleration coefficients. Also *w* in the velocity updating formula is called inertia weight and represents a factor for contribution of previous velocity of the particle.

Step 5. Steps 3 to 4 are repeated until a predetermined maximum iteration number (*maxt*) is reached.

Algorithm 3. Main Algorithm of the RANFIS

Step 1 The length of test set is determined and the time series is partitioned to two data sets as training set and test set.
Step 2 FCM method [7] is applied to training set and the rules are determined by using [8] method.
Step 3 Parameters of particle swarm optimization are determined and RANFIS is trained by using Algorithm 2.
Step 4 The forecasts are calculated for test set by using Algorithm 1 and trained RANFIS.

APPLICATION

For the evaluation of the forecasting performance of the proposed method, five real time series were analyzed. These five time series include daily data from 2009-2013 Istanbul Stock Exchange Market BIST100 index. Test sets include 7 and 15 observations for each time series. The proposed method, autoregressive integrated moving average model (ARIMA), simple exponential smoothing (ES), ANFIS with subtractive clustering and modified ANFIS (MANFIS) [8] are applied to time series and results of ten applications are given in Table **1**. Five time series are listed below and their graphs are given in Figs. (**2**) - (**6**). In all these graphs in Figs. (**2**) - (**6**), vertical and horizontal axes represent observed values and date, respectively.

Series 1: BIST100 index values with 103 observations observed between 01.02.2009 and 05.29.2009 (Fig. **2**).
Series 2: BIST100 index values with 104 observations observed between 01.05.2010 and 05.31.2010 (Fig. **3**).
Series 3: BIST100 index values with 106 observations observed between 01.03.2011 and 05.31.2011 (Fig. **4**).
Series 4: BIST100 index values with 106 observations observed between 01.02.2012 and 05.31.2012 (Fig. **5**).
Series 5: BIST100 index values with 106 observations observed between

01.02.2013 and 05.31.2013 (Fig. **6**).

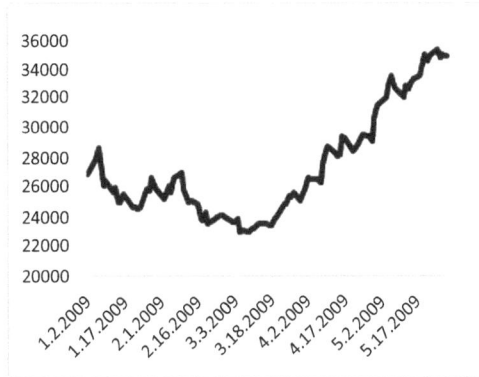

Fig. (2). Time series graph of Series 1.

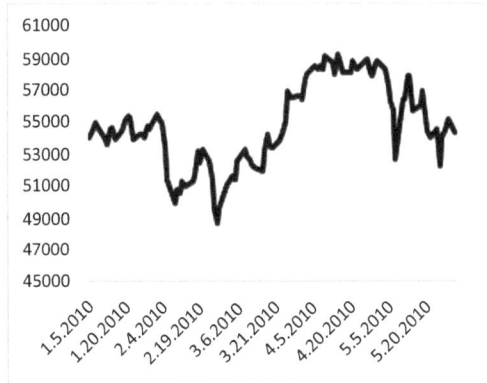

Fig. (3). Time series graph of Series 2.

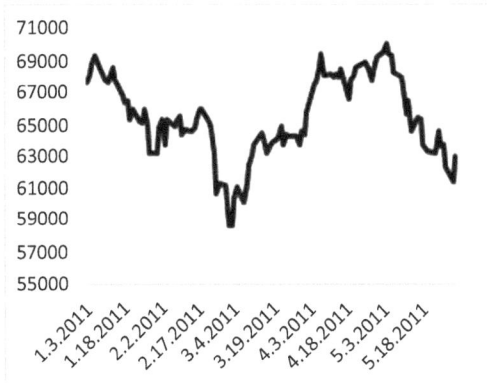

Fig. (4). Time series graph of Series 3.

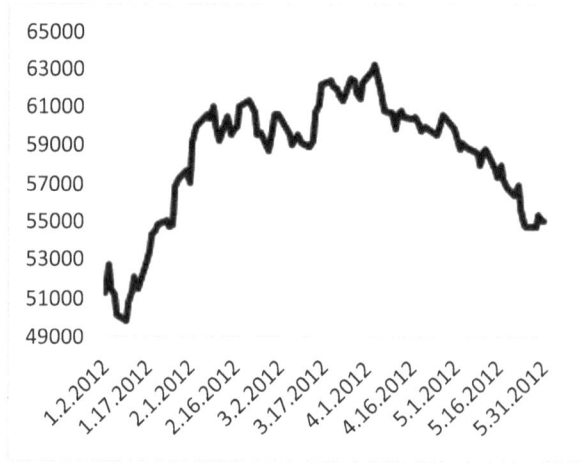

Fig. (5). Time series graph of Series 4.

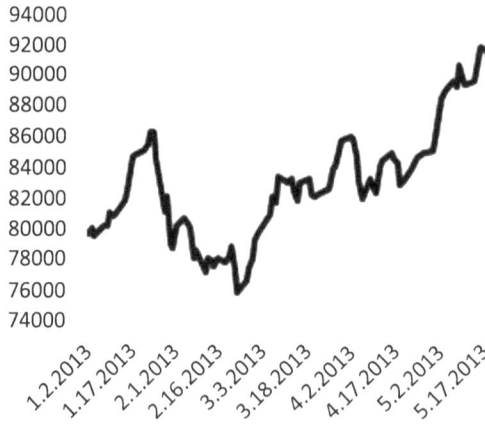

Fig. (6). Time series graph of Series 5.

Table 1. The RMSE Values Obtained From Test Sets For Ten applications.

Data Set	ARIMA	ES	ANFIS	MANFIS	RANFIS
2009/7	344	344	405.15	261	**171**
2009/15	540	540	647.28	503	**473**
2010/7	1221	1221	1141.3	1144	**905**
2010/15	1612	1612	2033	1303	**1138**
2011/7	1057	1057	1007	960	**941**
2011/15	1129	1129	1134	1009	**1002**
2012/7	651	651	634	634	678
2012/15	621	621	938	629	639

(Table 1) contd.....

Data Set	ARIMA	ES	ANFIS	MANFIS	RANFIS
2013/7	1361	1361	1477	1418	**1338**
2013/15	1268	1268	1413	1264	1298

CONCLUSION

ANFIS is the most preferred fuzzy inference system in the literature. ANFIS is a rule based system and it can be used to obtain forecast for time series. In the literature, classical time series models and artificial neural networks use errors of the system as inputs of system. These kind of systems are called recurrent systems. ANFIS is a feed forward system. The contribution of this paper is to propose a recurrent fuzzy inference system. In this study, a new recurrent ANFIS is introduced. In the RANFIS, lagged square of errors are inputs in right sides of the rules. In the new RANFIS, error terms don't affect the number of rules apart from other recurrent ANFIS. According to application results, RANFIS can produce accurate forecasts. RANFIS gave the best forecasting performance in 70% of all applications. In future studies, different recurrence types can be tried on the ANFIS.

CONFLICT OF INTEREST

The authors (editor) declares no conflict of interest, financial or otherwise.

ACKNOWLEDGEMENTS

Declared none.

REFERENCES

[1]   J. Zhang, and A.J. Morris, "Recurrent Neuro-Fuzzy Networks for Non-linear Processes Modelling", *IEEE Trans. Neural Netw.,* vol. 10, no. 2, pp. 313-326, 1999.
[http://dx.doi.org/10.1109/72.750562]

[2]   H. Tamura, K. Tanno, H. Tanaka, C. Vairappan, and Z. Tang, "Recurrent type ANFIS using local search Technique for time series prediction", *APCCAS IEEE Asia Pacific Conference,* 2008.
[http://dx.doi.org/10.1109/APCCAS.2008.4746039]

[3]   H. Yu, and Z. Tang, "Learning Recurrent ANFIS using Stochastic Pattern Search Method", *IJCSNS International Journal of Computer Science and Network Security,* vol. 10, no. 6, pp. 277-283, 2010.

[4]   N. Kasuan, N. Ismail, M.N. Taib, and M.H. Rahiman, "Recurrent Adaptive Neuro-Fuzzy Inference System for Steam Temperature Estimation in Distillation of Essential Oil Extraction Processes", *IEEE 7th International Colloquium on Signal Processing and Its Applications,* 2011.

[5]   M.S. Mahmud, and P. Meesad, "An Innovative recurrent error-based neuro-fuzzy system with momentum for stock price prediction", *Soft Computing,* vol. 20, no. 10, pp. 4173-4191, 2016.
[http://dx.doi.org/10.1007/s00500-015-1752-z]

[6]   J. Kennedy, and R. Eberhart, "Particle swarm optimization", *Proceedings of IEEE International Conference on Neural Networks,* pp. 1942-1948, 1995.
[http://dx.doi.org/10.1109/ICNN.1995.488968]

[7]   J.C. Bezdek, *Pattern recognition with fuzzy objective function algorithms.* Plenum Press: NY, 1981, pp. 65-80.
[http://dx.doi.org/10.1007/978-1-4757-0450-1]

[8]   E. Egrioglu, C. H. Aladag, U. Yolcu, and E. Bas, "A New Adaptive Network Based Fuzzy Inference System For Time Series Forecasting", *Aloy Journal of Soft Computing and Applications,* vol. 2, no. 1, pp. 25-32, 2014.

CHAPTER 10

A Hybrid Method for Forecasting of Fuzzy Time Series

Eren Bas[*]

Giresun University, Faculty of Arts and Sciences, Department of Statistics, Giresun, Turkey

Abstract: Fuzzy time series approaches have been used for real world time series contain uncertainty. When these approaches are used, it is not necessary to satisfy the assumptions needed for conventional time series methods. Fuzzy time series methods are composed of three phases which are fuzzification, determination of fuzzy relations, and defuzzification. Artificial intelligence algorithms are widely employed in these phases. Genetic algorithm and differential evolution algorithm are one of the most popular artificial intelligence algorithms. Besides, the hybrid algorithms by obtaining the composed of some artificial intelligence algorithms have been frequently used in the literature. In this paper, a hybrid method composed by genetic algorithm and differential evolution algorithms is proposed to find the optimal interval lengths. The hybrid method proposed in this paper has been applied to Canadian lynx data and its superior forecasting performance was shown when compared with those obtained by other techniques suggested in the literature.

Keywords: Differential evolution algorithm, Forecasting, Fuzzy time series, Genetic algorithm, Hybrid method, Mutation operator.

INTRODUCTION

Fuzzy time series methods do not need the assumptions that classic time series approaches have such as autoregressive moving average models [1]. The interest to fuzzy time series methods has been gradually increased because of its advantages. Artificial intelligence techniques have been used in almost all areas in the literature [2 - 8]. The fuzzy set theory was firstly touched on the study of [9]. Fuzzy time series were first introduced by Song and Chissom [10 - 12]. Fuzzy time series methods consist of three stages. These stages are; fuzzification, determination of fuzzy relations and defuzzification. In the literature, there are many studies for each stage to improve the forecasting performance.

In fuzzification stage, Song and Chissom [10 - 12] and Chen [13, 14] determined

[*] **Corresponding author Eren Bas:** Giresun University, Faculty of Arts and Science, Department of Statistics, 28200, Giresun, Turkey; E-mail: eren.bas@giresun.edu.tr

Cagdas Hakan Aladag (Ed.)

the interval lengths as arbitrarily. Huarng [15] suggested two new techniques for finding the intervals. Egrioglu and coworkers [16, 17] suggested finding the intervals as an optimization problem. Chen and Chung [18], Lee and coworkers [19] and Uslu and coworkers [20] used dynamic interval lengths instead of fixed interval lengths *via* genetic algorithm (*GA*), Bas and coworkers [21, 22] and [23 - 25] used particle swarm optimization. Besides, Cheng and coworkers [26], Li and coworkers [27] and Cagcag Yolcu [28] used fuzzy c-means clustering methods. And also, Egrioglu and coworkers [29] and Alpaslan and Cagcag [30] used Gustafson-Kessel fuzzy clustering in this stage.

In determination of fuzzy relations stage, Song and Chissom [10 - 12] used matrix operations. Chen [13] and some others used to prefer a fuzzy logic relations group table instead of complex matrix operations. And also [31 - 34], used artificial neural networks to determine fuzzy relations.

In defuzzification stage, the centroid method proposed in [13, 15, 35] is the most using method and also, adaptive expectation method proposed in [26, 36] was also used in this stage.

In this study, a hybrid method (*HM*) which composed of *GA* and differential evolution algorithm (*DEA*) together is proposed to find the optimal interval lengths. Besides, more realistic results are obtained by using this *HM* and the intervals are determined by avoiding subjective judgments. The other parts of the paper are the methods used in study, the proposed method, application and conclusions parts.

THE METHODS USED IN THIS STUDY

Fuzzy Time Series

The definition of fuzzy time series was firstly introduced in [10, 11].

Let U be the universe of discourse, where $U = \{u_1, u_2, \ldots, u_n\}$. A fuzzy set A_i of U can be defined as,

$$A_i = \frac{\mu_{A_i}(u_1)}{u_1} + \frac{\mu_{A_i}(u_2)}{u_2} + \cdots + \frac{\mu_{A_i}(u_n)}{u_n} \tag{1}$$

where μ_{A_i} is the membership function of the fuzzy set A_i and $\mu_{A_i} ; U \rightarrow [0,1]$. Besides, $\mu_{A_i}(u_1)$, $j = 1,2,\ldots,n$ denotes is a generic element of fuzzy set A_i; $\mu_{A_i}(u_j)$, is the degree of belongingness of u_1 to A_i; $\mu_{A_i}(u_j) \in [0,1]$.

Definition 1. Fuzzy time series Let $Y(t)$ ($t = \ldots,0,1,2\ldots$) a subset of real numbers, be the universe of discourse by which fuzzy sets $f_i(t)$ are defined. If $F(t)$ is a collection of $f_1(t), f_2(t), \ldots$ then $F(t)$ is called a fuzzy time series defined on $Y(t)$.

Definition 2. Fuzzy time series relationships assume that $F(t)$ is caused only by $F(t-1)$, then the relationship can be expressed as: $F(t) = F(t-1)*R(t, t-1)$, which is the fuzzy relationship between $F(t)$ and $F(t-1)$, where * represents as an operator. To sum up, let $F(t-1) = A_i$ and $F(t) = A_j$. The fuzzy logical relationship between $F(t)$ and $F(t-1)$ can be denoted as $A_i \rightarrow A_j$ where A_i (current state) refers to the left-hand side and A_j (next state) refers to the right-hand side of the fuzzy logical relationship. Furthermore, these fuzzy logical relationships can be grouped to establish different fuzzy relationship.

Genetic Algorithm

GA was suggested by Holland [37]. Goldberg [38] made some contributions *GA* has been successfully solving many complex optimization problems. It does not require the specific mathematical analysis of optimization problems. A *GA* structure consists of chromosomes and these chromosomes are very important for genetic algorithm process. *GA* imitates the evolutionary process for solving the problems. *GA* has some parameters and these parameters are very important for the process of *GA*. These parameters are population size, crossover rate, mutation rate and also if necessary the other genetic operators can be used such as repairing operator. There are many studies in the literature about the parameters of genetic algorithms [39, 40]. studied on these parameters and suggested some parameter values about these parameters. The *GA* search optimal solution with many chromosomes. In one chromosome, there are many gens. The gens are decision variables when *GA* is used for optimization. Therefore, researchers firstly should pay attention to the coding type of the gens.

Differential Evolution Algorithm

DEA was proposed by Price and Storn [41]. *DEA* is a heuristic algorithm based on the population. Some operators of *DEA* are similar with *GA* such as crossover and mutation operators. But the using of these operators is different from each other. These operators are applied to every chromosome in the population different from *GA*. And also, there are many mathematical operations in *DEA* process. These mathematical operations are used to create new chromosomes in *DEA* process. At the end of the process of *DEA*, some nominee chromosomes are obtained and they transfer to the new generation according to the fitness function value. *DEA* and its operators discussed in more detail in the section of proposed method and also those who want more information can look the study of Price and Storn [41].

PROPOSED METHOD

As it is well known that all stages of the fuzzy time series approaches influences very much intensively on the forecasting performance of the model and researchers have used different techniques to make a contribution to each stages. In recent years, artificial intelligence algorithms have been used in fuzzification stage by the researchers. And also, the hybrid algorithms by obtaining the composed of some artificial intelligence algorithms have been frequently used in the literature.

From this point of view, in this paper, *GA* and *DEA* are used together by considering the features of their operators, in the stage of fuzzification as a structure of *HM*. The aim of the paper is to take the forecasting performance a step further and to make decisions without subjectively. From this point of view, more dynamic interval lengths can be obtained without subjective judgments.

The novelties of the paper are given below:

Using dynamic length of intervals improves the forecasting performance and gives more reliable results. There is no need to subjective judgments to define the interval lengths by using the *HM* algorithm. This hybrid algorithm was proposed in Bas and coworkers [42]:

Step 1. In this step, the universe of discourse and the how many intervals are used are determined

Firstly, the surplus of U and the real universe are determined as (D_{min}) and (D_{max}) and

$$U = [D_{\min}, D_{\max}] \qquad (2)$$

In a chromosome structure of *HM*, there are (m-1) genes. The representation of a chromosome is given in Fig. (**1**). In this Fig, $x_i (i = 1, 2, ..., m-1)$; shows each gene of a chromosome.

x_1	x_2	$\cdots$	x_{m-1}

Fig. (1). The structure of a chromosome in *HM.*.

Then the margins of all intervals are given by

$$u_1 = [D_{\min}, x_1], u_m = [x_{m-1}, D_{\max}], \qquad (3)$$

Step 2. The parameters used in the structure of *HM* and the initial

The parameters are; the total chromosome number in a population (*tcn*), the ratio of crossing over (*rco*) ($0 < rco \le 1$), the total chromosome number to be eliminate in natural selection (*ecn*) and the number of repetitions *(rpt)*.

After determining the parameters, *tcn* is generated by using uniform distribution between (D_{min}, D_{max}). Every *chromosome* has (m−1) genes. The values in each chromosome have ascending order.

Step 3. Root Mean Square Error (*RMSE*) is determined as an evaluation function for each chromosome. It is calculated from the steps 3.1 to 3.5.

Step 3.1. *Determine the fuzzy sets* (A_i).

$$A_i = {a_{i1}}\Big/{u_1} + {a_{i2}}\Big/{u_2} + \cdots + {a_{im}}\Big/{u_m} \quad i = 1, 2, \cdots, m \tag{4}$$

In Equation 4, *m* is the number of intervals. And also, each fuzzy set is created by using m intervals based on gene values.

where a_{ik} is the degree of membership and it is given in Equation 5

$$a_{ik} = \begin{cases} 1 & k = i \\ 0.5 & k = i-1, i+1, \quad i = 1, 2, \cdots, m \\ 0 & otherwise \end{cases} \tag{5}$$

The fuzzy sets are created from the observations of a crisp time series and the interval corresponds to the related observation has the highest membership values.

Step 3.2. Determine Fuzzy Relations and Fuzzy Relation Group Tables

If there is a relation like $F(t-1) = A_i$ and $F(t) = A_j$ for any time *t* is observed, this fuzzy relation is given by $A_i \rightarrow A_j$. For the whole series, if the relation $F(t-1) = A_i$ and $F(t) = A_k$ for any time *t* are obtained, the fuzzy relation is given by $A_i \rightarrow A_j$, A_k. The number of times of the fuzzy relations such as $A_i \rightarrow A_j$ occurs was saved, as the weight w_j.

Step 3.3. Determine fuzzy forecasts.

The fuzzy forecasts are determined *via* fuzzy relation table touch on Step 3.2. For instance; if $F(t-1) = A_i$ and if $A_i \rightarrow A_j$ in the fuzzy relation table, the fuzzy forecast is A_j. If $F(t-1) = A_i$ and there is a relation such as $A_i \rightarrow A_j, A_k$ in the fuzzy relation table, the fuzzy forecast is A_j, A_k. If $F(t-1) = A_i$ and $A_i \rightarrow Empty$ in the fuzzy logic relation table, the fuzzy forecast is A_i.

Step 3.4. Defuzzification of the fuzzy forecasts

At the end of the creation of fuzzy relation table, the weights w_j are determined. The weights mean that how many relations are repeated how many times.

For instance, if $F(t-1) = A_i$ and there is a relation $A_i \rightarrow A_j$ in the fuzzy relation table, the defuzzified forecast is m_j. m_j is the midpoint of the subinterval u_j of A_j with the highest membership degree. In other words, how many times the relation is repeated in the table is not considered.

If $F(t-1) = A_i$ and there is a relation $A_i \rightarrow A_j, A_k$ and w_j symbolizes how many times $A_i \rightarrow A_j$ is repeated in the whole time series and w_k symbolizes how many times $A_i \rightarrow A_j$ is repeated, the defuzzified forecast is calculated as in Equation 6

$$\hat{x}_t = \frac{w_j m_j + w_k m_k}{w_j + w_k} \tag{6}$$

If $F(t-1) = A_i$ and there is a relation $A_i \rightarrow Empty$ in the fuzzy relation table, the defuzzified forecast is m_i,

Step 3.5. *RMSE* is calculated as in Equation 7.

$$RMSE = \sqrt{\frac{\sum_{t=1}^{n}(x_t - \hat{x}_t)^2}{n}} \tag{7}$$

In this Equation, x_t is the real time series and $\hat{x}_t$ are defuzzified forecasts respectively.

Step 4. Keep the chromosome with minimum evaluation function in the population

Step 5. Determine and use the *HM* operators.

The operators of *HM* are; natural selection, crossover and mutation.

Step 5.1. Firstly, *ecn* in the population is determined and the number of *ecn* is sifted. The number of *ecn* is determined by considering the evaluation function. High *RMSE* values are undesired situation for the population. In other words, the high *RMSE* values mean bad solution sets. *ecn* is sifted from ordered *RMSE* values from highest to lowest. After they sifted, new chromosomes as *ecn* as are generated randomly applied in Step 2 and put to the population.

Step 5.2. In order to apply crossover operator, firstly a random number is generated between 0 and 1 by using uniform distribution. If this random number is not bigger than *rco*, crossover operator can be applied to the population. Before applying crossover operator, two chromosomes are selected randomly from the population and then a crossover point is determined randomly. According to the cross over point, the genes are interchanged.

Step 5.3. The last operator of *HM* is mutation operator. Firstly, generally, the first chromosome is selected and it is called as related chromosome. Except for this chromosome, three different chromosomes are determined randomly. The difference of the first two of these chromosomes is taken and created a new chromosome called difference vector. After, each gene of this vector is multiplied by *F* parameter value and created a new chromosome called weighted difference vector. (This value is taken as 0.8 in accordance with the literature). Finally, this chromosome is summed with the third chromosome determined randomly at the beginning and created a new chromosome called total vector. This new chromosome, called the weighted difference vector is summed with the third chromosome to obtain the total vector. Afterwards, a mutation ratio (*mtnr*) is selected in accordance with *DEA* literature and a random number is generated between 0 and 1 by using uniform distribution for each gene. If this random number is not bigger than *mtnr*, the gene is taken from total vector. If it is bigger than *mtnr*, the gene is taken from related chromosome. Considering the gene transfer, a new chromosome called nominee chromosome is created and the fitness function value of this chromosome is calculated. The fitness function values of the related chromosome and nominee chromosome are calculated. The chromosome has smallest *RMSE* is transferred to the population.

This process is applied every chromosomes in the population, respectively.

Step 6. Steps 3 to 5 are repeated as the number of repetition.

APPLICATION

In the application section of the paper, Canadian lynx data was analyzed to prove the performance of the proposed *HM*. In the analysis of the data, mean absolute percent error (*MAPE*) criteria given in Equation 8 is also used for the comparison of the proposed method with some other methods in fuzzy time series literature.

$$MAPE = \frac{1}{n} \sum_{t=1}^{n} \frac{\left| x_t - \hat{x}_t \right|}{x_t} * 100 \tag{8}$$

The parameters of *HM* are:

- *tcn* was tried between 20 and 100 with an increasing 10.
- *rco* was tried between 0.1 and 1 with an increasing 0.1.
- *ecn* was tried as 7, 10, 13, 17, 20, 23, 26, 30, and 33.
- For all possible cases given above, *HM* was run 100 times in MATLAB.
- *m* was tried from 5 to 20.

Considering all parameters, it can be obtained 1440 different solution combinations. The best solution among all these solutions is the parameters *(m, tcn, rco, ecn)* with the smallest *RMSE*.

Analysis of Canadian Lynx Data

Canadian Lynx data consisting of the set of annual numbers of lynx trappings in the Mackenzie River District of North-West Canada with observations for the period from 1821 to 1934 was analyzed both the proposed method and the methods in papers [10, 13, 43] and Egrioglu [44] which are in fuzzy time series literature. The logarithm (to the base 10) of the data was used in the analysis. The last 14 observations were separated as test data. In Table **1**, the forecasts both obtained from some other methods in fuzzy time series literature and proposed *HM* were given together. Besides, *RMSE* and *MAPE* values are calculated by using these forecasts for all methods without excepting the proposed method. At the end of the analysis, the best result was obtained with *m=11, tcn=60, rco=0.05, and ecn=20*

Table 1. Comparison of the results for test set.

Test Data	[13]	[10]	[43]	[44]	[42]
5.4300	5.1770	4.9608	4.6365	6.2024	5.4955
5.9890	5.8256	5.9877	5.2851	6.2024	5.9590
7.0317	6.2039	6.2116	6.5823	6.2024	6.3343
7.7965	7.0507	7.2309	8.5281	7.0461	7.4757
8.1814	7.1228	7.2309	8.5281	7.9342	7.4757
7.9845	7.6849	7.2309	7.2309	7.9342	7.4757
7.3376	7.1228	7.2309	6.5823	7.9342	7.4757
6.2710	7.1660	7.2309	6.5823	7.0461	7.4757

(Table 1) contd.....

Test Data	[13]	[10]	[43]	[44]	[42]
6.1841	6.1138	6.2580	6.5823	6.2024	6.3343
6.4953	6.2039	6.2116	6.5823	6.1581	6.3343
6.9078	6.1138	6.2580	6.5823	6.2024	6.3343
7.3715	7.0507	7.2309	7.2309	6.2024	7.4757
7.8850	7.1660	7.2309	7.2309	7.9342	7.4757
8.1304	7.1228	7.2309	7.8795	7.9342	7.4757
RMSE	0.6407	0.6202	0.5349	0.5932	0.5214
MAPE	0.0755	0.0726	0.0692	0.0699	0.0569

It is clearly seen in Table **1**, the proposed *HM* has the best forecasting performance in terms of both *RMSE* and *MAPE* criteria when compared with other techniques suggested in the literature.

CONCLUSIONS

When examined the fuzzy time series literature, it seen that the researchers have different contributions with using different methods for each stage of fuzzy time series. In recent years, different artificial intelligent techniques such as *GA* have been used by researchers in each stage of fuzzy time series techniques and it is seen that these techniques give successful results on the forecasting performance of the model.

From this point of view, in this study, a *HM* was used in fuzzification stage. This study contributes the fuzzy time series literature by using *DEA* and *GA* in the same algorithm called *HM* and the superior forecasting performance of the proposed method is shown by analyzing a real life time series data.

In the next studies, different contributions for any stage of fuzzy time series can be made and different hybrid methods can be used.

CONFLICT OF INTEREST

The author (editor) declares no conflict of interest, financial or otherwise.

ACKNOWLEDGEMENTS

Declared none.

REFERENCES

[1]   G.E. Box, and G.M. Jenkins, *Time Series Analysis: Forecasting and Control.* Holdan-Day: San Francisco, CA, 1976.

[2]   S.A. Khan, and A.P. Engelbrecht, "A fuzzy particle swarm optimization algorithm for computer communication network topology design", *Appl. Intell.,* vol. 36, no. 1, pp. 161-177, 2012.

[3]   J.G. Kang, S. Kim, S.Y. An, and S.Y. Oh, "A new approach to simultaneous localization and map building with implicit model learning using neuro evolutionary optimization", *Appl. Intell.,* vol. 36, no. 1, pp. 242-269, 2012.

[4]   W.T. Pan, "The use of genetic programming for the construction of a financial management model in an enterprise", *Appl. Intell.,* vol. 36, no. 2, pp. 271-279, 2012.

[5]   M.H. Kang, H.R. Choi, H.S. Kim, and B.J. Park, "Development of a maritime transportation planning support system for car carriers based on genetic algorithm", *Appl. Intell.,* vol. 36, no. 3, pp. 585-604, 2012.

[6]   Y.M. Ali, "Psychological model of particle swarm optimization based multiple emotions", *Appl. Intell.,* vol. 36, no. 3, pp. 649-663, 2012.

[7]   K.S. Shin, Y.S. Jeong, and M.K. Jeoung, "A two-leveled symbiotic evolutionary algorithm for clustering problems", *Appl. Intell.,* vol. 36, no. 4, pp. 788-799, 2012.

[8]   R. Abbasian, and M. Mouhoub, "A hierarchical parallel genetic approach for the graph coloring problem", *Appl. Intell.,* vol. 39, no. 3, pp. 510-528, 2013.

[9]   L.A. Zadeh, "Fuzzy Sets", *Inf. Control,* vol. 8, pp. 338-353, 1965.

[10]   Q. Song, and B.S. Chissom, "Fuzzy time series and its models", *Fuzzy Sets Syst.,* vol. 54, pp. 269-277, 1993.

[11]   Q. Song, and B.S. Chissom, "Forecasting enrollments with fuzzy time series- Part I", *Fuzzy Sets Syst.,* vol. 54, pp. 1-10, 1993.

[12]   Q. Song, and B.S. Chissom, "Forecasting enrollments with fuzzy time series- Part II", *Fuzzy Sets Syst.,* vol. 62, pp. 1-8, 1994.

[13]   S.M. Chen, "Forecasting enrollments based on fuzzy time-series", *Fuzzy Sets Syst.,* vol. 81, pp. 311-319, 1996.

[14]   S.M. Chen, "Forecasting enrollments based on high order fuzzy time series", *Cybern. Syst.,* vol. 33, pp. 1-16, 2002.

[15]   K. Huarng, "Effective length of intervals to improve forecasting in fuzzy time-series", *Fuzzy Sets Syst.,* vol. 123, pp. 387-394, 2001.

[16]   E. Egrioglu, C.H. Aladag, U. Yolcu, V.R. Uslu, and M.A. Basaran, "Finding an optimal interval length in high order fuzzy time series", *Expert Syst. Appl.,* vol. 37, pp. 5052-5055, 2010.

[17]   E. Egrioglu, C.H. Aladag, M.A. Basaran, V.R. Uslu, and U. Yolcu, "A new approach based on the optimization of the length of intervals in fuzzy time series", *J. Intell. Fuzzy Syst.,* vol. 22, pp. 15-19, 2011.

[18]   S.M. Chen, and N.Y. Chung, "Forecasting enrolments using high order fuzzy time series and genetic algorithms", *Int. J. Intell. Syst.,* vol. 21, pp. 485-501, 2006.

[19]   L.W. Lee, L.H. Wang, S.M. Chen, and Y.H. Leu, "Handling forecasting problems based on two factor high-order fuzzy time series", *IEEE Trans. Fuzzy Syst.,* vol. 14, no. 3, pp. 468-477, 2006.

[20]   V.R. Uslu, E. Bas, U. Yolcu, and E. Egrioglu, "A fuzzy time series approach based on weights determined by the number of recurrences of fuzzy relations", *Swarm Evol. Comput.,* vol. 15, pp. 19-26, 2014.

[21]   E. Bas, V.R. Uslu, U. Yolcu, and E. Egrioglu, "A fuzzy time series analysis approach by using differential evolution algorithm based on the number of recurrences of fuzzy relations", *American Journal of Intelligent Systems,* vol. 3, no. 2, pp. 75-82, 2013.

[22]   E. Bas, V.R. Uslu, U. Yolcu, and E. Egrioglu, "A fuzzy time series approach using de/best/1 mutation

strategy of differential evolution algorithm", *Aloy Journal of Soft Computing and Applications,* vol. 2, no. 2, pp. 60-69, 2014.

[23] F.P. Fu, K. Chi, W.G. Che, and Q.J. Zhao, "High-order difference heuristic model of fuzzy time series based on particle swarm optimization and information entropy for stock markets", *International Conference on Computer Design and Applications,* 2010.

[24] Y.L. Huang, S.J. Horng, T.W. Kao, R.S. Run, J.L. Lai, R.J. Chen, I.H. Kuo, and M.K. Khan, "An improved forecasting model based on the weighted fuzzy relationship matrix combined with a PSO adaptation for enrollments", *Int. J. Innov. Comput., Inf. Control,* vol. 7, no. 7, pp. 4027-4046, 2011.

[25] I-H. Kuo, S-J. Horng, T-W. Kao, T-L. Lin, C-L. Lee, and Y. Pan, "An improved method for forecasting enrollments based on fuzzy time series and particle swarm optimization", *Expert Syst. Appl.,* vol. 36, pp. 6108-6117, 2009.

[26] C.H. Cheng, T.L. Chen, H.J. Teoh, and C.H. Chiang, "Fuzzy time series based on adaptive expectation model for TAIEX forecasting", *Expert Syst. Appl.,* vol. 34, pp. 1126-1132, 2008.

[27] S.T. Li, Y.C. Cheng, and S.Y. Lin, "FCM-based deterministic forecasting model for fuzzy time series", *Comput. Math. Appl.,* vol. 56, pp. 3052-3063, 2008.

[28] O. Cagcag Yolcu, "A hybrid fuzzy time series approach based on fuzzy clustering and artificial neural network with single multiplicative neuron model", In: *Mathematical Problems in Engineering,* 2013. Article ID 560472, 9 pages

[29] E. Egrioglu, C.H. Aladag, U. Yolcu, V.R. Uslu, and N.A. Erilli, "Fuzzy time series forecasting method based on Gustafson–Kessel fuzzy clustering", *Expert Syst. Appl.,* vol. 38, pp. 10355-10357, 2011.

[30] F. Alpaslan, and O. Cagcag, "A seasonal fuzzy time series forecasting method based on Gustafson-Kessel fuzzy clustering", *Journal of Social and Economic Statistics,* vol. 1, pp. 1-13, 2012.

[31] C.H. Aladag, M.A. Basaran, E. Egrioglu, U. Yolcu, and V.R. Uslu, "Forecasting in high order fuzzy time series by using neural networks to define fuzzy relations", *Expert Syst. Appl.,* vol. 36, pp. 4228-4423, 2009.

[32] E. Egrioglu, C.H. Aladag, U. Yolcu, V.R. Uslu, and M.A. Basaran, "A new approach based on artificial neural networks for high order multivariate fuzzy time series", *Expert Syst. Appl.,* vol. 36, pp. 10589-10594, 2009.

[33] E. Egrioglu, C.H. Aladag, U. Yolcu, M.A. Basaran, and V.R. Uslu, "A new hybrid approach based on SARIMA and partial high order bivariate fuzzy time series forecasting model", *Expert Syst. Appl.,* vol. 36, pp. 7424-7434, 2009.

[34] E. Egrioglu, V.R. Uslu, U. Yolcu, M.A. Basaran, and C.H. Aladag, "A new approach based on artificial neural networks for high order bivariate fuzzy time series", In: *Applications of Soft Computing, AISC 58 Springer-Verlag Berlin Heidelberg,* J. Mehnen, Ed., pp. 265-273, 2009.

[35] K. Huarng, and T.H. Yu, "Ratio-based lengths of intervals to improve fuzzy time series forecasting", *IEEE Trans. Syst. Man Cybern. B Cybern.,* vol. 36, pp. 328-340, 2006.

[36] C.H. Aladag, U. Yolcu, and E. Egrioglu, "A high order fuzzy time series forecasting model based on adaptive expectation and artificial neural networks", *Math. Comput. Simul.,* vol. 81, pp. 875-882, 2010.

[37] J.H. Holland, *Adaptation in Natural and Artificial Systems.* University of Michigan Press: Ann Arbor, 1975.

[38] D.E. Goldberg, *Optimal Initial Population Size for Binary-Coded Genetic Algorithms.* Department of Engineering Mechanics, University of Alabama: Alabama, 1985.

[39] K.A. De Jong, Analysis of the behavior of a class of genetic adaptive systems, Department of Computer and Communication Science, University of Michigan: Ann Arhor, MI, 1975.

[40] J.D. Schaffer, R.A. Caruana, L.J. Eshelman, and R. Das, "A study of control parameters affecting on-

line performance of genetic algorithms for function optimization", *Proc. 3rd Int. Conference on Genetic Algorithms and their Applications,* 1989.

[41]   R. Storn, and K. Price, "Differential evolution: a simple and efficient adaptive scheme for global optimization over continuous spaces", *Technical Report TR-95-012, International Computer Science Institute, Berkeley,* 1995.

[42]   E. Bas, V.R. Uslu, U. Yolcu, and E. Egrioglu, "A modified genetic algorithm for fuzzy time series to find the optimal interval lengths", *Appl. Intell.,* vol. 41, no. 2, pp. 453-463, 2014.

[43]   C.H. Aladag, "Using multiplicative neuron model to establish fuzzy logic relationships", *Expert Syst. Appl.,* vol. 40, pp. 850-853, 2013.

[44]   E. Egrioglu, "PSO-based high order time invariant fuzzy time series method: Application to stock exchange data", *Econ. Model.,* vol. 38, pp. 633-639, 2014.

SUBJECT INDEX

A

Activation function 83, 99, 100, 101, 131
Actual TAIEX 67, 71, 72
Adaptive expectation model, employed 112
Adaptive networks fuzzy inference systems 156
Adaptive neuro fuzzy inference system (ANFIS) 145, 156, 157, 160, 163
Additive neuron models 83
AFER model validation 73
Algorithm 20, 26, 28, 29, 30, 69, 80, 81, 86, 87, 88, 96, 102, 146, 157, 158, 159, 160, 165, 168, 173
  hybrid 165, 168
Analysis methods 93, 94, 113, 128
Analysis process 93, 95, 96, 104, 107, 127, 128, 133
ANFIS, inputs of 156
ANFIS architecture 156
ANN and Fuzzy Approaches 94
ANN models 140
Antecedents 38, 39, 40, 42, 43
Approaches 10, 24, 26, 32 40, 43, 93, 94, 95, 100, 107, 112, 128, 140
  based 10
  dependent 43
  designed 40
  first-order 145
  hybrid 26, 32
  independent 43
  new 100
  non-probabilistic 24, 93, 107
  non-stochastic 94
  novel 112
  probabilistic 24
  second 42
  stochastic 24
  subjective 140
  systematic 95, 128
ARIMA and SARIMA models 80
Artificial intelligence algorithms 165, 168

Artificial neural network model, new 77
Artificial neural network models 81
Artificial neural networks (ANNs) 24, 25, 28, 76, 77, 80, 81, 82, 83, 84, 85, 95, 107, 127, 128, 133, 140, 144, 145, 151, 166
Artificial neuron models 81
Assumptions, strict 94, 127
Australian beer consumption time series 148, 149
Autoregressive 26, 30, 80, 78, 148, 151, 160, 165
  seasonal 30, 80, 148, 151
Autoregressive models 24, 26, 32, 84, 85, 145, 156
  linear 24, 85
  nonlinear 84, 156
Autoregressive Part (ARP) 28
Average forecasting error rate (AFERs) 48, 57, 58, 59, 60, 66, 67, 68, 71, 72

B

Basic defuzzification distribution (BADD) 43, 73
Best forecasting model 3
Best forecasting performance 24, 156, 163, 173
Biological neuron model, real 81
Bonds exchange market of İstanbul 4, 10, 20, 111, 113, 117, 124
BP algorithm 81, 82

C

Calculation of forecasted values 116, 120
Chromosome number, total 169
Chromosomes 98, 104, 167, 171
  best 98, 104
  new 98, 167, 171
  nominee 167, 171
  third 171
Cloud density 40, 50, 51, 52, 53, 54, 55

www.ingramcontent.com/pod-product-compliance
Lightning Source LLC
Chambersburg PA
CBHW041702210326
41598CB00007B/499